Tissue Engineering Intelligence Unit 4

Support of the Acutely Failing Liver

Second Edition

Achilles A. Demetriou, M.D., Ph.D.
Frederick D. Watanabe, M.D.
Cedars-Sinai Medical Center
Los Angeles, California, U.S.A.

CRC Press
Taylor & Francis Group
Boca Raton London New York

CRC Press is an imprint of the
Taylor & Francis Group, an **informa** business

Support of the Acutely Failing Liver

Second Edition

Tissue Engineering Intelligence Unit

CRC Press
Taylor & Francis Group
6000 Broken Sound Parkway NW, Suite 300
Boca Raton, FL 33487-2742

First issued in paperback 2019

ISBN-13: 978-1-58706-001-4 (pbk)

Library of Congress Cataloging-in-Publication Data

CIP applied for, but not available at time of publication.

Visit the Taylor & Francis Web site at
http://www.taylorandfrancis.com

and the CRC Press Web site at
http://www.crcpress.com

CONTENTS

EDITORS

Achilles A. Demetriou, M.D., Ph.D.
Department of Surgery
Liver Support Unit
Esther and Mark Schulman Chair
in Surgery and Transplantation Medicine
Cedars-Sinai Medical Center
Los Angeles, California, U.S.A.
Chapters 5, 7-12

Frederick D. Watanabe, M.D.
Liver Transplant
Liver Support Unit
Cedars-Sinai Medical Center
Los Angeles, California, U.S.A.
Chapters 5, 8-10

CONTRIBUTORS

Nikolaos Arkadopoulos
Liver Support Research Laboratory
Cedars-Sinai Medical Center
Los Angeles, California, U.S.A.
Chapters 4, 9, 12

Walid S. Arnaout
Liver Transplantation
Center for Liver Diseases
and Transplantation
Cedars-Sinai Medical Center
Los Angeles, California, U.S.A.
Chapters 6, 7, 11

Steven D. Colquhoun
Liver Transplantation
Center for Liver Diseases
and Transplantation
Cedars-Sinai Medical Center
Los Angeles, California, U.S.A.
Chapter 3

Caroline A. Connelly
Center for Liver Diseases
and Transplantation
Cedars-Sinai Medical Center
Los Angeles, California, U.S.A.
Chapter 3

Olivier Detry
Liver Support Research Laboratory
Cedars-Sinai Medical Center
Los Angeles, California, U.S.A.
Chapter 4

Susumu Eguchi
Liver Support Research Laboratory
Cedars-Sinai Medical Center
Los Angeles, California, U.S.A.
Chapter 12

Donna E. Goldman
Center for Liver Diseases
and Transplantation
Cedars-Sinai Medical Center
Los Angeles, California, U.S.A.
Chapter 2

Elaine A. Kahaku
Liver Support Research Laboratory
Cedars-Sinai Medical Center
Los Angeles, California, U.S.A.
Chapters 1, 5, 8, 9

Theodore Khalili
Liver Support Research Laboratory
Cedars-Sinai Medical Center
Los Angeles, California, U.S.A.
Chapter 8

Helene Lilja
Liver Support Research Laboratory
Cedars-Sinai Medical Center
Los Angeles, California, U.S.A.
Chapters 7, 11, 12

Jody E. Margulies
Liver Support Research Laboratory
Cedars-Sinai Medical Center
Los Angeles, California, U.S.A.
Chapter 4

David A. Moscioni
Liver Support Research Laboratory
Cedars-Sinai Medical Center
Los Angeles, California, U.S.A.
Chapter 12

Anthony Navarro
Liver Support Research Laboratory
Cedars-Sinai Medical Center
Los Angeles, California, U.S.A.
Chapter 8

David Reybould
Cedars-Sinai Medical Center
Los Angeles, California, U.S.A.
Chapter 6

Raoul J. Rosenthal
Liver Support Research Laboratory
Cedars-Sinai Medical Center
Los Angeles, California, U.S.A.
Chapter 12

Jack Rozga
Surgical Research, CSMC
Cedars-Sinai Medical Center
Los Angeles, California, U.S.A.
Chapters 7, 9, 11, 12

Kyung S. Suh
Liver Support Research Laboratory
Cedars-Sinai Medical Center
Los Angeles, California, U.S.A.
Chapter 7

Paul Ting
Liver Support Research Laboratory
Cedars-Sinai Medical Center
Los Angeles, California, U.S.A.
Chapter 8

Graham M. Woolf
Cedars-Sinai Medical Center
Los Angeles, California, U.S.A.
Chapter2

PREFACE

Severe liver dysfunction results in massive physiologic derangement and high mortality. A concerted multidisciplinary effort is needed to support patients with severe acute liver failure (SALF) and treat them appropriately. This requires mobilization of significant resources. There is a need to put in place clinical teams to provide comprehensive diagnostic and therapeutic plans and to critically assess emerging technologic advances. In addition to clinical expertise, there is a need for an appropriate scholarly, scientific environment that can allow meaningful experimental studies to be carried out to advance knowledge in the field. Five years ago, at Cedars-Sinai Medical Center, we established the Liver Support Unit (LSU) to meet these needs.

In this monograph, members of the LSU present the most current understanding of the pathophysiology of liver failure and how its various forms and manifestations are classified, and summarize the state of the art in the diagnosis and management of the disease. In addition, basic laboratory as well as clinical experimental work carried out in the LSU is presented, with particular emphasis on the use of liver support systems in an attempt to "bridge" patients with severe forms of liver failure to either transplantation or liver repair and regeneration.

This work is dedicated to the clinical and research fellows trained at the LSU in recognition of their dedication, creativity and hard work, both at the bedside and in the laboratory; to the Critical Care and General Clinical Research Center nurses; and to the Perfusion Team members at our institution for their support and commitment to achieving the best possible clinical care.

Achilles A. Demetriou
Frederick D. Watanabe
Los Angeles, California, U.S.A.

CHAPTER 1

The Liver: A Brief Historical Perspective

Elaine Kahaku

The United Nations has recommended standardized dating using BCE, for 'before the common era', instead of BC and CE, for 'common era', instead of AD.

Throughout history, man has striven to understand the liver's function and disorders. Early representations of the liver were seen in the form of clay models of sheep livers from the Assyro-Babylonian era (3000-2000 BCE).[1] Ancient cultures believed that the liver was responsible for generating heat and blood, qualities essential for life and soul. But it was not until the Classical Age of Greece when early beliefs, ideas and concepts of pathogenesis of liver disease were developed. Greek physicians regarded the liver, and particularly the portal system, as the starting point of many disorders. Hippocrates (460-375 BCE) believed that the liver secreted two humors, yellow and black bile, out of the four which constituted the essential components of bodily processes.[2] The correct proportion of humors assured good health, and humor disproportion was responsible for development of disease. Excess yellow bile was believed to be associated with the qualities of warmth and dryness. Yellow bile induced acute disease and acted on the brain to cause restlessness and delirium. In contrast, black bile, with cold and dry qualities, produced chronic disease, debility and other nervous system disorders.

Plato (428-347 BCE) believed that the liver was "the manager for the body and the origins of physical appetites" (Timaeus, 70-23 BCE).[3] Herophilus (335-280 BCE) and Erasistratus (310-250 BCE) were the first to give an accurate description of human liver anatomy and postulate the existence of an intrahepatic capillary bed and formation of bile by filtration.[3-5] But it was not until Galen (129-199 CE) that broader liver physiology principles were established. Galen believed that the liver retained a dominant role as the "seat of sanguification and the source of the veins".[6] He believed that the liver separated bile from blood and that it provided the means for chyle to be transformed into blood.[6] Galen's system of hepatic pathology and pathogenesis of liver disease prevailed for centuries.

In the late 15th century, William Harvey, in his *De Motu Cordis* (1628) correlated the function of the heart to blood flow and circulation.[7,8] Glisson (1654), a young colleague of Harvey, described the liver capsule and showed the close relationship between the portal venous hepatic and biliary systems.[8] Malpighi (1628-1694) believed that the liver was an intricate and complex gland with the bile duct as its excretory duct.[9] With the aid of a primitive microscope, Malpighi was among the first generation of histologists to recognize the lobular architecture of the liver.[10] But it was not until the advent of modern microscopes that histologists such as Dutrochet (1837), Henle (1837) and others identified and established the appearance of hepatocytes.[11] By the early part of the 19th century, physicians were able to diagnose liver disease based upon microscopic observations. Louis (1825)

Support of the Acutely Failing Liver, 2nd ed., edited by Achilles A. Demetriou and Frederick D. Watanabe.

and Addison (1836) were the first to describe fatty liver changes, and Kiernan (1833), Carswell and Hallmann differentiated cirrhosis of the liver from hepatic tumors.[12-16]

Alcoholic cirrhosis was first described by English clinicians in the eighteenth century after the "gin plague" but it was Matthew Baillie (1793) who, in his book *Morbid Anatomy*, described the association between cirrhosis ("common tubercle of the liver") with consumption of alcohol.[17,18] The term cirrhosis is derived from the Greek word "kirrhos", meaning orange-colored. Laënnec, (1819) in his book *De L'auscultation Médiate*, proposed the term cirrhosis to describe chronic diffuse fibrosis of the liver in alcoholics after noticing the yellowish color of granular deposits in the liver.[19] Bamberger (1864), and earlier Freichs (1861), differentiated the morphologic course of cirrhosis into two stages: an inflammatory stage, followed by the development of shrinking, nodular scar tissue.[20] Cirrhosis, as the clinical-pathologic syndrome we know today, was summarized by Mallory (1911) in his important paper "Cirrhosis of the Liver"; he expanded the term cirrhosis to indicate "a chronic, progressive, destructive lesion of the liver combined with reparative activity and contraction on the part of the connective tissues."[21] Mallory emphasized the key elements of the definition of cirrhosis, namely the presence of parenchymal destruction and scarring, as well as nodular parenchymal regeneration.[21]

"Epidemic jaundice", which was later replaced with "infective hepatitis" (England, 1939) or "infectious hepatitis" (USA, 1943) was the first term used to describe viral hepatitis.[22] During the 19th century, epidemic jaundice flourished among soldiers during wartime, drawing the attention of many clinicians. During the Civil War (1861-1865), approximately 42,000 cases of jaundice were reported, with 161 deaths.[23] Epidemic jaundice was also observed in both civilian and military populations during the Franco-Prussian War of 1870. The Japanese navy experienced vast numbers of hepatitis cases in the war against Russia (1904-1905), as did the German navy in World War I. Using clinical and epidemiological data, Louis Kelsch (1841-1911) showed the relationship of epidemic jaundice and cirrhosis. In 1942, after inoculation of yellow fever vaccine containing human serum, 28,505 American soldiers in Africa developed jaundice, with 62 deaths. It was not until World War II that research in viral hepatitis accelerated.[24] Yoshibumi and Shigemoto (Japan, 1941) were the first to establish experimental transmission of hepatitis, and Blumberg and colleagues (1965) described the serologic marker for the hepatitis B virus.[25,26]

Early therapy of liver disease dates back to the Classical Greek period. Hippocrates (500 BCE) and later, Celsus (100 CE) and Aretaeus (200 CE), recommended evacuation of hepatic abscesses by surgically opening the abdomen with a knife or cautery.[27] Thyme, castor oil, sea onion and calomel were ingested to promote diuresis in patients with ascites and fluid overload.[28] Fresh raw pigeon liver was routinely fed to cirrhotics.[29-31] To release body fluid, cathartics, bloodletting and application of leeches were utilized.[29-31] Decompensated cirrhosis was mainly treated by paracentesis for relief of ascites. Although asepsis was unknown, there were surprisingly few cases of peritonitis, as illustrated in the works of such leading clinicians as Bamberger (1822-1888), Andral and Frerichs, who seldom reported complications.[32-34]

In the eighteenth and nineteenth centuries, physicians began to have a more clear understanding of the various disease entities of the liver. The era of specific therapy emerged, with the gradual abandonment of heroic phlebotomy, massive applications of leeches and severe purgation. Later, clinicians such as von Liebermeister in 1892 and Minkowski in 1907 observed retardation of the cirrhotic process after withdrawal of alcohol.[28] Other physicians, such as Becquerel in 1840, went further in their dietary restrictions, prohibiting coffee and spices.[28] The belief was that these substances caused liver congestion and, either by precipitation of intravascular fibrin or overloading of the metabolic functions, led to hepatitis and cirrhosis.

Early recognition of hepatic encephalopathy can be traced as far back as Hippocrates during the fifth century, when he observed symptoms of jaundice associated with episodic delirium among cirrhotic patients.[35] Shakespeare, through the creation of Sir Andrew Aguecheck (*Twelfth Night*, 1603) described symptoms of hepatic coma after ingestion of large amounts of protein. Ballonius (1671) was the first to describe hepatic coma in an alcoholic man with jaundice, and Morgagni (1761) was the first to clinically describe and demonstrate the autopsy finding of a fatty liver.[30,36] Bright (1836) reported several cases of acute liver failure and Herxheimer (1966) introduced a classification of acute, subacute and chronic forms of hepatitis.[37,38] Trey and Davidson (1970) were the first to attempt to classify the stages of hepatic encephalopathy; this classification was further expanded by Bernuau and others in 1986.[39,40]

It is clear that the role of the liver as a regulator of multiple bodily processes has been recognized for centuries. Modern medical science has defined the physiologic and molecular mechanisms of liver function and dysfunction. However, in spite of major advances in our understanding of liver disease, major gaps remain. To improve our knowledge, we need to integrate the observations of many past investigators with current knowledge and develop creative ideas for future basic and clinical experimentation.

References

1. Jastrow M. The liver in antiquity and the beginnings of anatomy. College of Physicians Philadelphia 1907; 29:117-138.
2. Rothschuh KE. History of physiology. Transl GB Risse Huntington, NY: Krieger 1973; 155-160.
3. Knell AJ. Liver function and failure: The evolution of liver physiology. J Royal Coll Physicians of London 1980; 14:205-08.
4. Dobson JF. Herophilus of Alexandria. Proc Roy Soc Med 1925; 18:19-32.
5. Dobson JF. Erasistratus. Proc Roy Soc Med 1927; 20:825-832.
6. French RK. The thorax in history. 2. Hellenistic experiment and human dissection. Thorax 1978; 33:153-166.
7. Harvey W. The anatomical lectures: Prelectiones anatomie universalis. In: Whitteridge G, transl. Edinburgh: Livingstone, 1964:103, 125-7, 131, 137-9, 141, 143, 147-149.
8. Chen TSN, Chen PSY. William Harvey as hepatologist. Am J Gastroenterology 1988; 83:1274-77.
9. Chen TS, Chen PS. Understanding the liver; a history. Westport, Conn: Greenwood Press 1984:27-31.
10. Chen TS. Anatomic Foundation; Understanding the liver; a history. Westport, Conn: Greenwood Press 1984:28-29.
11. Baker JR. The cell theory: A restatement, history and critique. Part 2. Quart J Microscop Sci 1949; 90:87-108.
12. Louis PCA. Recherches anatomiques, pathologiques et therapeutiques sur la phthisie. Gabon, Paris, 1825.
13. Addison TA. Observations on fatty degeneration of the liver. Guy's Hospital Report 1836; 1:476.
14. Kiernan F. The anatomy and physiology of the liver. Phil Trans Roy Soc London 1833; 123:711-770.
15. Carswell R. Pathological anatomy. Illustrations of the elementary forms of diseases. Chapter on Atrophy. London: Longman 1838.
16. Hallmann E. De cirrhosi hepatis. Diss Inaug. Berolini, 1839; 27. Cited in Davidson Charles. Cirrhosis of the Liver. Chicago: Year Book Publishers, Sept 1957.
17. Thurn R J. The gin plague. Minn Med 1978; 61:241-243.
18. Baillie M. The morbid anatomy of some of the most important parts of the human body. Reprinted in AE Rhodin. The influence of Matthew Baillieís morbid anatomy. Springfield, Ill: C.C. Thomas 1973:72.

19. Laennec RTH. De l'auscultation médiate. Vol 1. Paris: Brosson et Chaudé 1819:368-369.
20. Frerichs FT. Diseases of the Liver, Vol 2, Guy's Hosp Rep 1836; 1:24.
21. Mallory FB. Cirrhosis of the liver. Five different types of lesions from which it may arise. Bull Johns Hopkins Hosp 1911; 22:69-75.
22. Chen TS, Chen PS. Disorders of the liver; understanding the liver; a history. Westport, Conn: Greenwood Press 1984:113-137.
23. Chen TS, Chen PS. Disorders of the liver; understanding the liver; a history. Westport, Conn: Greenwood Press 1984:117.
24. Editorial. Jaundice following yellow fever vaccination. J Amer Med Assoc 1942; 119:1110.
25. Yoshibumi H, Shigemoto T. Human experiment with epidemic jaundice. Acta Pediat Jap 1941; 47:975.
26. Blumberg BS et al. A "new" antigen in leukemia sera. J Amer Med Assoc 1965; 191:541-546.
27. Celsus AC. De medicina, Vol 1, 413. Transl WG Spencer 1960. Classical Library, London: Heinemann.
28. Franken FH, Falk H. The history of liver therapy. In: Schaffner F, Sherlock S, Leevy C. eds. The Liver and Its Diseases. New York: Intercontinental Medical Book Corporation, 1974:117-25.
29. Broussais FJ. Histoire des phlegmasies ou inflammations chroniques. Paris, 1808.
30. Fraken FH. Die Leber und ihre Krankeiten Zweihundert Jahre Hepatologie. Enke, Stuttgart, 1968.
31. Mani N. Die historischen Grundlagen der Leberforschung. I. Teil, Schwabe, Basel: Stuttgart, 1959.
32. Bamberger H v. Die Krakheiten der Leber. In: Handb. d. spez. Pathologic und Therapie, redig: v R Virchow, 2. Aufl., Erlangen, 1864.
33. Andral G. Clinique médicale, Observations sur les maladies de l'abdomen. Paris 1834 Dtsch. Ubers.: Med. Klinik, 2. Bd. 2. Theil. Quedlinburg u. Leizzig, 1843.
34. Frerich FT. Klinik der Leberkrankheiten. Braunschweig, 2 Bde. 1858-1861.
35. Hippocrates, The sacred disease. In: Adams F. (transl.), The genuine works of Hippocrates. New York: W. Wood, 1929:344.
36. LePage, J. Rozga, Historical Perspective; Support of the Acutely Failing Liver. Austin, TX 1994.
37. Bright. Bright's cases of jaundice. Guy's Hosp 1836; 1:621
38. Herxheimer G. Über akute gelbe Leberatrophie und verwandte Veränderungen II. Beitr Pathol Anat 1924; 72:349-417.
39. Trey C, Davidson LS. The management of fulminant hepatic failure. In: Popper H, Schaffner F. eds. Progress in Liver Disease. New York: Grune and Stratton 1970; 282-8.
40. Bernuau J, Rueff B, Benhamou J-P. Fulminant and subfulminant liver failure: Definitions and causes. Sem Liver Dis 1986; 6:97-106.

CHAPTER 2

Acute Liver Failure: Definitions and Etiology

Donna E. Goldman, Graham M. Woolf

In the United States, there are approximately 2000 cases/year of acute hepatic failure (AHF) in adults and children.[1] The term "fulminant hepatitis" was introduced by Lucke and Mallory in 1948 to describe posttransfusion hepatitis in soldiers dying within 9 days of the onset of their illness.[2] Since that time, the criteria for defining fulminant hepatic failure (FHF) have evolved due to improved understanding of the etiology, prognosis and complications of the disease. However, in spite of our improved understanding of the disease process, the definition and classification of FHF continue to remain confusing and often controversial.

Definitions: Types of Hepatic Failure

Acute Hepatic Failure

Acute hepatic failure (AHF) is defined as severe liver dysfunction with prothrombin time prolongation and a coagulation factor V level less than 50% in the absence of encephalopathy (Table 2.1).[3] Although this initial stage is usually associated with jaundice and elevated serum transaminase levels, its definition is based only on significant changes in coagulation parameters. Patients with plasma factor V levels greater than 50% rarely develop encephalopathy and have better prognosis.[4]

Fulminant and Subfulminant Hepatic Failure

Acute hepatic failure can progress to fulminant hepatic failure (FHF) with the onset of encephalopathy.[3] The presence of encephalopathy is mandatory for the definition of FHF.

Trey and Davidson Classification

FHF was first defined in 1970 (by Trey and Davidson) as the appearance of encephalopathy less than eight weeks after the onset of symptoms in a patient with no known prior liver disease (Table 2.1).[5] It is now recognized that FHF can also occur in patients with underlying chronic liver disease.

Bernuau Classification

Bernuau et al in 1986 defined FHF as AHF complicated by encephalopathy occurring less than two weeks after the onset of jaundice.[3] Bernuau et al recognized that prognosis and clinical findings were different in those patients who developed encephalopathy later than 2 weeks. They subsequently introduced the term subfulminant hepatic failure (SFHF) to describe AHF complicated by encephalopathy occurring 2-12 weeks after the onset of

Support of the Acutely Failing Liver, 2nd ed., edited by Achilles A. Demetriou and Frederick D. Watanabe.

Table 2.1. Definitions of liver failure

	Acute Hepatic Failure	Fulminant Hepatic Failure	Subfulminant Hepatic Failure	Hyperacute Liver Failure	Acute Liver Failure	Subacute Liver Failure	Late Onset Hepatic Failure
Encephalopathy	No	Yes	Yes	Yes	Yes	Yes	Yes
Time from onset of jaundice to encephalopathy	None	< 8 weeks or < 2 weeks	2-12 weeks	0-7 days	8-28 days	5-12 weeks	2-6 months
Cerebral edema	No	Frequent	Rare	Frequent	Frequent	Rare	Rare
Ascites	None	Rare	Frequent	Rare	Rare	Frequent	Frequent
Prognosis	Good	Poor to Fair	Poor	Fair	Poor	Poor	Poor
References	3	3,5	3	6	6	6	5

All patients with prothrombin time and factor V less than 50% have severe liver dysfunction.

jaundice (Table 2.1).[3] History of prior liver disease does not preclude a diagnosis of FHF or SFHF with these definitions.

FHF is associated with a mortality rate greater than 80% with medical management alone;[6] the rate is higher in comatose compared to noncomatose patients.[3] FHF is more commonly associated with cerebral edema and a hepatitis A or B viral etiology than SFHF. The latter is less frequently associated with cerebral edema and more frequently associated with non-A and non-B hepatitis (NANBH) and drug toxicity. Survivors of FHF can recover fully and do not develop chronic liver disease.

Late Onset Hepatic Failure (LOHF)

The above definitions did not take into account patients with prolonged illness prior to the onset of encephalopathy. Various terms have been used to describe this disease entity, including subacute or subchronic atrophy of the liver, subacute hepatitis, subacute (fatal) hepatitis and subacute hepatic necrosis. More recently, the term LOHF has been introduced by Gimson to describe patients who develop encephalopathy 2-6 months after the onset of liver dysfunction (Table 2.1).[7] Generally, in contrast to patients with FHF, patients with LOHF have an older median age (45 years vs. 30 years), increased incidence of NANBH (compared to hepatitis A or B), increased severe cholestasis and ascites and decreased incidence of cerebral edema. Elevated immunoglobulin G levels and anti-smooth muscle and anti-nuclear antibodies suggest a possible overlap between LOHF and chronic active hepatitis. LOHF mortality is high (81%) and survivors often develop chronic liver disease and post-necrotic cirrhosis, in contrast to survivors of FHF.

O'Grady Classification

The above definitions have attempted to clarify the relationship between prognosis and time of onset of encephalopathy. However, after a detailed review of their FHF patients, O'Grady et al proposed a new terminology (Table 2.1).[8] Hyperacute liver failure was defined as the development of encephalopathy within 7 days of the appearance of jaundice. These patients had a fair prognosis with a survival rate of 36%, despite a 69% incidence of cerebral edema. Etiology in this group included all cases of paracetamol overdose and many hepatitis A and B virus infections. Patients with acute liver failure developed encephalopathy 8-28 days after the onset of liver dysfunction. Even though the incidence of cerebral edema (56%) was similar to that seen in hyperacute liver failure, the survival rate was only 7%. Indeterminate etiology accounted for 46% of these cases. Subacute hepatic failure occurred when encephalopathy developed 5-12 weeks after the onset of jaundice. Although the incidence of cerebral edema was only 14% in these patients, prognosis was poor, with a survival rate of 14%. Etiology was indeterminate in 86% of these patients. This new proposed classification included patients with a previous history of asymptomatic chronic liver disease.

In an accompanying letter to O'Grady's publication, Bernuau and Benhamou disagreed with the new terminology.[9] In their own series of FHF patients, they found little difference in survival between patients with hyperacute and those with acute liver failure, perhaps due to differences in underlying etiology.[3] Bernuau and Benhamou were also critical of defining "acute liver failure" based upon the time interval between jaundice and encephalopathy rather than representing the entire syndrome, including coagulation abnormalities. They felt that a more accurate determination of prognosis could be obtained based on etiology, age and prothrombin or factor V levels, indices more reliably determined than the onset of jaundice and appearance of encephalopathy.

FHF terminology remains controversial, although there is agreement on certain aspects. In the Bernuau et al and O'Grady et al classifications, encephalopathy is an important element.[3,8] Although the length of time between the development of encephalopathy and the onset of jaundice differs, it is apparent that patients with a shorter course (less than 2 weeks) have better prognosis than those with a prolonged course (greater than 2 weeks). Hepatitis A and B viruses and acetaminophen toxicity occur more frequently in cases with a shorter course, whereas an indeterminate etiology is the primary cause in cases with a prolonged course. Further work is required to identify the underlying etiologies in indeterminate cases.

Severe Acute Liver Failure (SALF)

In several parts of this monograph, we will use the term SALF to describe patients with severe acute liver failure and massive liver necrosis on histologic examination, irrespective of the length of the interval between the onset of encephalopathy and the development of jaundice. In addition, we will use the terms FHF and SFHF as defined above.[3]

Etiology

There are four major etiologies for hepatic failure (Tables 2.2 and 2.3):

1. Viral;
2. Drug-induced;
3. Toxin-induced;
4. Miscellaneous.

In the United States, Spain, France and Britain the majority of cases are secondary to acute viral hepatitis, particularly hepatitis B. Hepatotoxic drugs are the second most common cause in the United States and Spain, whereas toxins (hydrocarbons and mushroom toxins) are more frequent in France.[3,10-12] Acetaminophen and halothane were the more common hepatotoxic drugs causing SALF in Britain. The etiology is unknown or indeterminate in up to 50% of cases. In a review of 2,550 cases of FHF, survival ranged from 14-35%, except in acetaminophen toxicity, where survival was as high as 50%.[13]

Viral Causes of Liver Failure

Viral Hepatitis

Hepatitis A (HAV)

The incidence of fulminant and subfulminant HAV is very low and occurs in less than 0.01% of all HAV infections.[3] Younger patients rarely develop hepatic failure. HAV accounted for less than 6% of FHF and SFHF cases due to viral hepatitis in the United States and France but 31% of the cases in Britain.[14-16] HAV infection results in FHF more frequently than SFHF.[17] Survival with medical therapy is relatively high, ranging between 40 and 60%.[16,18] Relapse of HAV occurs in 10% of patients, usually within 2-3 months after clinical improvement from the initial illness.[19] Relapse is recognized by an increase in transaminases and bilirubin, with reappearance of the virus in the stool. If encephalopathy occurs during this period, the outcome is usually fatal.[20]

Hepatitis B (HBV)

Fulminant and subfulminant HBV accounts for less than 1% of HBV infections, but is the most common cause of fulminant viral hepatitis.[21-23] In the United States and France, the prevalence of HBV in patients with FHF is 60 to 65%, but in Britain it is only 25%.[3,14,16] As is the case in HAV infection, HBV results more commonly in FHF than SFHF.[3] Hepatitis B surface antigen (HBsAg) and HBV DNA may be absent in some cases of HBV FHF.[24-26] Studies have reported negative HBsAg tests in 2 out of 17 patients with fulminant HBV and negative HBV DNA tests in 58 of 64 such patients.[5,24] Serum HBsAg concentrations may be

Table 2.2. Causes of acute, fulminant and subfulminant hepatic failure

1) Infectious
   Viruses: Hepatitis A, B, C, D and E
   Hepatitis non-A, non-E (indeterminate)
   Coxsackie B, Herpes simplex, Varicella, Cytomegalovirus, Measles, Epstein-Barr virus, Lassa, Adenovirus, Yellow fever, Giant-cell hepatitis, Ebola, Dengue parvovirus B19, Togavirus, Marburg arbovirus group B
   Bacteria: Q fever
   Parasites: Ameba

2) Drugs (Table 2.3)

3) Toxins
   Mushrooms: *Amanita phalloides, Amanita verna, Amanita virosa,* Lepiota species, Bacillus cereus
   Hydrocarbons: Carbon tetrachloride, Trichloroethylene, 2-nitropropane, Chloroform, Monochlorobenzene
   Copper
   Aflatoxin
   Yellow phosphorus

4) Miscellaneous Conditions
   Wilson's disease
   Acute fatty liver of pregnancy
   Reye's sydrome
   Hypoxic liver cell necrosis
   Hypo or Hyperthermia
   Budd-Chiari syndrome
   Veno-occlusive disease of the liver
   Autoimmune hepatitis
   Massive malignant infiltration of the liver
   Partial hepatectomy
   Liver transplantation
   Post-jejunoileal bypass
   Galactosemia
   Hereditary fructose intolerance
   Tyrosinemia
   Erythropoietic protoporphyria
   Irradiation
   Alpha-1-antitrypsin deficiency
   Niemann-Pick II (C)
   Neonatal hemochromatosis
   Cardiac tamponade
   Right ventricular failure
   Circulatory shock
   Tuberculosis
   Leptospirosis
   Malaria
   Q fever
   Amebiasis

Table 2.3. Drugs causing acute, fulminant or subfulminant hepatic failure

Nonsteroidal anti-inflammatory drugs		
	Acetaminophen	Salicylates
	Piroxicam	Pirprofen
	Ibuprofen	Indomethacin
	Naproxen	Diclofenac
	Oxaprozin	Sulindac
	Etodolac	Felbamate
Neuropsychiatric		
	Valproic acid	Diphenylhydantoin
	Carbamazepine	Felbamate
	Iproclozide	Prochloroperazine
	Imipramine	Phenobarbital
	Desipramine	Amitriptyline
	MAO inhibitors	
Antibiotics		
	Tetracycline	Sulfasalazine
	Isoniazid (with or without Rifampicin)	Ketoconozole
	2′3′-Dideoxyinosine (ddI)	Nitrofurantoin
	Sulfonamides	Amodiaquine
	Griseofulvin	Cotrimoxazole
	Amoxicillin plus Clavulinic acid	Dapsone
Anesthetics		
	Halothane	Enflurane
	Isoflurane	
Herbal medicines		
	Callilepis laureola (atractyloside)	Chaparral leaf
	Neurelax	Pyrrolizidine alkaloids
Hormonal drugs		
	Propylthiouracil	Flutamide
Hallucinogenic drugs		
	Cocaine	Phencyclidine (PCP)
	3,4-Methylenedioxymethamphetamine ("ecstasy")	
Cardiovascular drugs		
	Lisinopril	Methyldopa
	Hydralazine	Ecarazine hydrochloride
	Nicotinic acid	Lisinopril
	Labetalol	Amiodarone
Miscellaneous		
	Disulfiram	Cyproterone
	Fialuridinine (FIAU)	Omeprazole
	Interferon alpha	

significantly lower, and IgM antibody to hepatitis B core antigen (IgM anti-HBc) levels higher, in patients with FHF than those with non-FHF.[24] Wright et al evaluated the contribution of HBV infection in patients presumed to have FHF due to NANBH.[26] Although HBsAg and HBV DNA were absent in all sera, polymerase chain reaction (PCR) detected HBV DNA in the livers of 6 of 12 patients (50%) prior to undergoing liver transplantation. The clinical importance of this observation was underscored by the detection of serum HBsAg and HBV DNA post-liver transplantation in 3 and 5 patients respectively. Thus, 7 of

12 patients initially diagnosed with FHF due to NANBH had evidence of HBV infection. These findings indicate that in patients with FHF, an enhanced immune response prevents further HBV replication and results in more rapid clearance of HBsAg. The survival rate in patients positive for HBsAg on presentation (17%) is much lower than that of patients who are HBsAg negative (47%).[3] Clearances of HBsAg and HBV DNA result in better survival rates as well as decreased incidence of recurrence after emergency liver transplantation.[27-30] In chronic carriers, reactivation of HBV is recognized by the reappearance of IgM anti-HBc and HBV DNA.[31] HBV reactivation with encephalopathy is usually characterized by a subfulminant course.[32]

Hepatitis C (HCV)

FHF of indeterminate etiology is often attributed to NANBH.[3] Recently, the ability to diagnose HCV infection has markedly improved with the detection of anti-HCV by ELISA II, recombinant immunoblot assay (RIBA) II and HCV RNA by PCR.[33-35] The extent of the contribution of HCV infection to the indeterminate group remains controversial. In contrast to hepatitis A and B virus infections, SFHF is more common than FHF in hepatitis C patients.[3] Investigators in several centers were unable to detect HCV in patients with FHF.[36-39] Wright et al found no evidence of anti-HCV or HCV RNA by PCR in 15 patients with fulminant non-A, non-B hepatitis.[36] Feray et al and Sallie et al confirmed these findings in 53 patients.[37,39] Liang et al detected HCV RNA by PCR in only 2 of 17 patients who were negative for anti-HCV by ELISA II.[38] In contrast to these findings, two centers in Japan diagnosed HCV infection (ELISA II and PCR) in 3 of 7 and 10 of 17 patients with FHF, respectively.[40-42] Recently, our center reported the detection of HCV RNA by PCR in 9 of 15 (60%) patients with an indeterminate etiology for FHF.[43] A well documented case of post-transfusion FHF due to HCV has been reported. The patient became comatose 2 weeks after the onset of jaundice and required liver transplantation. Stored serum from the recipient revealed the presence of anti-HCV and HCV RNA. The transfused unit of donated blood was also positive for anti-HCV.[44]

Recent reports indicate that HCV may also be a cofactor in FHF due to HBV. Feray et al found that 8 of 17 HBsAg positive patients were also positive for HCV RNA.[37] Of these 8 patients, 5 were IgM anti-HBc positive and 3 were negative. HCV was only present in cases with an established etiology. In our institution, HCV RNA was found in 2 out of 7 patients with fulminant hepatitis B.[43] Wright et al suggest that variability in HCV results among centers may be due to differences in HCV strains, heterogeneous patient populations, use of different primers and protocols for HCV RNA detection and a superimposed cause for FHF in chronic anti-HCV positive patients.[45]

Hepatitis D (HDV)

Hepatitis D virus (delta agent) is a defective virus that replicates only in the presence of HBsAg, which is used as the envelope protein. HDV infection is present in 30-70% of parenteral drug abusers and is found in less than 1% of homosexual men.[23,46] HDV RNA is only detected in 10% of patients with fulminant hepatitis D.[47] HDV usually follows a subfulminant course.[4] HDV can occur either as a co-infection with HBV or as a super-infection in patients with chronic HBV.[48] HDV co-infection in an HBV patient increases the risk of FHF, but decreases its associated mortality (52% vs. 73%).[49,50] Of those patients with fulminant HDV, co-infection occurs more often (53-77%) than super-infection (23-47%).[50,51] FHF due to HDV super-infection has a higher mortality rate (72% vs. 52%) and more often predisposes to chronic liver disease (54% vs. 31%) than FHF due to HDV co-infection.[49,52,53] At least one case of FHF due to co-infection has been reported in a patient with AIDS.[54]

Hepatitis E (HEV)

HEV is a water-borne, epidemic form of NANBH.[55] HEV infection is enterally transmitted and self-limited, with no progression to chronicity. Epidemics have occurred in India, China, Africa and Mexico.[56-60] Acute HEV infection has been documented in the United States among travelers returning from endemic areas.[61,62] Acute HEV infection can be detected with a fluorescent antibody-blocking assay, anti-HEV immunoassays and PCR for HEV RNA.[63-65] Mortality from acute liver failure due to HEV is greater in females than males and is highest among infected pregnant women.[57] The increased HEV-associated mortality in pregnant women is thought to be related to host factors such as malnutrition or hormonal imbalance.[66] Recent reports from India indicate HEV as the cause of FHF in more than 75% of patients.[67] A report from Britain revealed that of 30 patients with an indeterminate cause of FHF, 5 had a positive PCR for HEV.[39] In contrast, an American study showed no evidence for HEV using PCR and ELISA in 17 patients with FHF of unknown etiology.[38]

Hepatitis Non-A, Non-B, Non-C (NANBNC, Indeterminate)

Despite the availability of advanced serological testing, there are many cases of FHF and SFHF with an indeterminate etiology.[3,10-12] These patients are placed into a NANBNC category, implying a viral etiology. A more accurate designation for this category is "indeterminate" since the etiology is unknown and may or may not be due to an undiagnosed viral hepatitis.

Other Viral Causes of Liver Failure

The herpes virus family rarely causes FHF or SFHF. There have been several reports of Herpes simplex virus types 1 and 2 producing massive hepatic necrosis, usually in patients who are pregnant or immunocompromised.[68-71] Reports of FHF with cytomegalovirus and Epstein-Barr virus infections have been controversial due to lack of availability of serological confirmation and/or exclusion of other etiologies.[72,73] Disseminated adenovirus can occur in immunosuppressed patients and presents with fever, pneumonia and liver dysfunction.[74] Syncytial giant cell hepatitis with intracytoplasmic structures resembling paramyxoviral nucleocapsids has also been reported to cause FHF.[75] A recently discovered RNA virus, hepatitis G (HGV), has been evaluated as a potential cause of FHF. In two studies, 50% of Japanese (3 of 6) and German (11 of 22) patients with FHF tested positive for HGV RNA by PCR.[76,77] Six of the 11 German patients had a primary diagnosis of fulminant hepatitis B, but their clinical courses were similar to those of patients with HGV alone.[77] At this time, there is no evidence that HGV causes FHF.

Drug-Induced Hepatotoxicity

Many drugs are known to be hepatotoxic.[78,79] Drug toxicity accounts for 15% of FHF and SFHF cases and runs a subfulminant course in 70% of patients.[4] Toxicity is usually unpredictable. Drug ingestion results in liver injury in less than 1% of patients, with 20% developing FHF or SFHF (Table 2.3).[3,80] Risk of either FHF or SFHF increases with an increase in total dose, simultaneous ingestion of other drugs which induce or inhibit hepatic enzymes and continuation of drug administration after the onset of liver disease.[3] Acetaminophen-induced FHF is associated with better prognosis than FHF caused by nonacetaminophen drugs.[4] N-acetylcysteine is the antidote for acetaminophen toxicity and ideally should be used early in the course of illness, although it may still be of benefit if used after the development of encephalopathy and coagulopathy.[81,82] Chronic alcohol ingestion potentiates acetaminophen toxicity due to glutathione depletion.[83,84] Prognosis is poor with acetaminophen-induced FHF if either the plasma pH is less than 7.3 in the presence of any grade of encephalopathy, or prothrombin time is greater than 50 seconds (INR>3.5) with a creatinine of less than 300 mmol/l in the presence of grade III or IV encephalopathy.[13]

Patients with nonacetaminophen drug-induced FHF have a poorer prognosis.[68] Severe liver failure has been reported in patients taking isoniazid.[85] Increased risk of toxicity occurs in those who chronically ingest alcohol.[86] Patients who ingest isoniazid with the enzyme inducer rifampin, have an increased risk of developing FHF. The interval between drug administration and onset of liver disease is shorter with rifampin.[87,88] Psychotropic drugs such as the monoamine oxidase inhibitors and tricyclic antidepressants have been reported to cause FHF, particularly when used in conjunction with hepatic enzyme inducers.[89,90] Nonsteroidal anti-inflammatory drugs (NSAIDs) can be hepatotoxic and several cases of FHF have been reported in patients using piroxicam, pirprofen, ibuprofen and indomethacin.[91-94] The mechanism of hepatocellular injury induced by these drugs is thought to be idiosyncratic rather than intrinsic toxicity.[95] Halothane-induced FHF occurs within 2 weeks of general anesthesia and has a mortality rate of greater than 90%.[96,97] Repeated administration increases the risk of FHF and reduces the interval between initial administration and onset of symptoms. Cocaine-induced FHF can occur and is histologically manifested as zonal necrosis and microvesicular steatosis.[98] 2',3'-dideoxyinosine (ddI) has been reported to be hepatotoxic and can lead to fulminant hepatic failure in AIDS patients, particularly if they have underlying liver disease.[99]

Toxins

Toxins account for less than 2% of FHF or SFHF.[3] Amanita mushroom poisoning and industrial hydrocarbons are involved in the majority of cases. Mushroom toxicity, caused by *Amanita phalloides verna* and *virosa*, has been reported in Europe and the United States. The active agents, phallotoxins and amanatoxins, have an enterohepatic circulation and are not destroyed by cooking.[100] As in acetaminophen-induced FHF, liver damage from mushroom toxicity is delayed and is usually proceeded by an 1-4 day period of vomiting and diarrhea. Amanita poisoning had a mortality rate of 22% in one series of 205 patients.[101] Mortality was associated with coma and coagulation abnormalities. Emergency liver transplant can be successful in patients with FHF due to amanita toxicity.[4]

Industrial hydrocarbons such as carbon tetrachloride and trichloroethylene are rare causes of FHF, and their use is now prohibited in industrial cleaners.[102] A nonchlorinated hydrocarbon, 2-nitropropane, found in coatings, inks, adhesives and paints, has also been reported to cause FHF following massive inhalation.[103] In third world countries, ingestion of aflatoxin and herbal medicines have been implicated as causes of FHF and SFHF.[104-106]

Miscellaneous Causes of Hepatic Failure

Wilson's Disease

Wilson's disease may present as FHF or SFHF with acute intravascular hemolysis and renal failure.[107,108] Family history of liver and neurological disease, Kayser-Fleischer rings and low serum ceruloplasmin levels help establish the diagnosis but may be absent.[109] Mortality is high with FHF, and therefore liver transplant is recommended.[110,111]

Acute Fatty Liver of Pregnancy

Acute fatty liver of pregnancy is a rare cause of FHF, accounting for 2% of cases in a large series.[112] Mortality rates are 85% for both mother and infant and 40% for the infant if the mother survives.[113] Delivery of the fetus results in regression of the microvesicular steatosis and abnormal liver tests.[114] The risk of FHF is increased with misdiagnosis and continuation of pregnancy.[115] Liver transplantation has been successfully performed for acute fatty liver of pregnancy associated with FHF.[116]

Reye's Syndrome

Reye's syndrome is another rare cause of FHF due to microvesicular fatty infiltration of the liver. This syndrome usually occurs in young children and may follow viral infections and salicylate ingestion.[117] Coma occurs in 25% of cases and is a poor prognostic sign. Liver transplantation results in improved survival.[118]

Hypoxic Liver Cell Necrosis

Ischemic necrosis of the liver may follow episodes of congestive heart failure, hypovolemic shock, hepatic artery thrombosis, myocardial infarction and cardiac tamponade.[119] The underlying mechanism involves a reduction of hepatic blood flow which results in centrilobular hepatic necrosis. In many cases the reduction in blood flow is transient, resulting in minimal liver injury. However, massive necrosis will result in FHF associated with marked serum transaminase level elevation.[120] Many of these patients are not suitable candidates for liver transplantation due to severe cardiovascular disease.

Hyperthermia

Exertional heatstroke had been reported to cause FHF associated with renal failure, hemolysis, and rhabdomyolysis.[121] The underlying mechanism is thought to be direct thermal injury to the tissues with hepatic ischemia. Volume replacement and cooling by mechanical or pharmacological means can be life-saving.[122]

Hepatic Vein Obstruction

Obstruction of the hepatic veins results in sinusoidal dilatation of the centrilobular region with ischemic necrosis of the surrounding parenchyma. If severe, this will progress to FHF.[123] Thrombosis of the large hepatic veins results in Budd-Chiari syndrome and is associated with myeloproliferative disorders, coagulation abnormalities and ingestion of oral contraceptives.[124] FHF due to Budd-Chiari syndrome is universally fatal and is often diagnosed at autopsy.[125] Liver transplantation or placement of a portal-systemic shunt results in a 70-80% survival.[126] Veno-occlusive disease of the liver affects the smaller centrilobular veins and can be caused by pyrrolizidine alkaloids, cancer chemotherapy and irradiation.[127,128] FHF has been reported in some patients with veno-occlusive disease following chemotherapy for malignancy.[129]

Autoimmune Hepatitis (AIH)

Acute liver failure due to autoimmune hepatitis can follow a subfulminant course over 1-3 months.[130] Patients are usually young females and may have type 2 AIH with liver/kidney microsomal-1 antibodies. FHF may develop in 10% of patients, with a 50% mortality.[131] Urgent liver transplantation is recommended for these patients.

Massive Malignant Infiltration of the Liver

Massive infiltration of the liver with either metastatic adenocarcinoma[1] or lymphoma can result in FHF or SFHF.[132-135] Mechanisms of liver failure include ischemia from sinusoidal invasion of the tumor cells and cytokine release.[133,136] In some cases of lymphoma, early chemotherapy can result in increased patient survival. Metastatic disease to the liver, lymphoma and primary hepatic tumors such as angiosarcoma and hemangioendothelioma are absolute contraindications for liver transplantation.[137,138]

Partial Hepatectomy

FHF can occur if more than 80% of a noncirrhotic liver is resected; a cirrhotic liver tolerates loss of parenchyma even less.[139,140] The risk of postoperative FHF in these patients depends on preoperative hepatic function, parenchymal disease activity and the extent of hepatic resection.[139,141]

Liver Transplantation

FHF can occur following liver transplantation due to primary graft nonfunction, hyperacute graft rejection, preservation liver injury and acute liver ischemia due to hepatic artery thrombosis.[142] Re-transplantation is required urgently if medical/surgical treatment is unsuccessful. Late onset FHF in the graft can occur because of recurrent viral hepatitis.[4]

Hepatic Failure in Infants and Children

FHF in infants is usually caused by an inborn error of metabolism such as galactosemia, hereditary fructose intolerance or tyrosemia.[143] Prognosis is better for children than adults, perhaps due to the greater regenerative capacity of the liver at a younger age. Other causes of FHF include herpes virus, echo virus, Epstein-Barr virus infections and hepatitis B.[27] In childhood, encephalopathy is the major complication and cause of death. In one series, 9 out of 31 (28%) patients survived. Mortality correlated well with the severity of encephalopathy but not with patient age or underlying etiology. Most cases were of indeterminate etiology, but five were due to paracetamol overdose, halothane and amanita mushroom poisoning.[144]

References

1. Berk PD, Popper H. Fulminant hepatic failure. Annotated abstracts of a workshop held at the National Institutes of Health, 1977. Am J Gastroenterol 1978; 69:349-400.
2. Lucke B, Mallory T. The fulminant form of epidemic hepatitis. Am J Pathol 1948; 22:867-946.
3. Bernuau J, Rueff B, Benhamou J-P. Fulminant and subfulminant liver failure: Definitions and causes. Sem Liver Dis 1986; 6:97-106.
4. Bernuau J, BenhamouJ-P. Fulminant and subfulminant hepatic failure. In: McIntyre N, Benhamou J-P, Bircher J et al, eds. Oxford Textbook of Clinical Hepatology. Oxford: Oxford Medical Publications 1991: 921-942.
5. Trey C, Davidson C. The management of fulminant hepatic failure. Prog Liver Dis 1970; 3:282-298.
6. Tygstrup N, Ranek L. Fulminant hepatic failure. Clin Gastroenterol 1981; 10:191-208.
7. Gimson AES, O'Grady J, Ede RJ et al. Late onset hepatic failure: Clinical, serological and histological features. Hepatology 1986; 6:288-294.
8. O'Grady J, Schalm SW, Williams R. Acute liver failure: Redefining the syndromes. Lancet 1993; 342:273-275.
9. Bernuau J, Benhamou J-P. Classifying acute liver failure. Lancet 1993; 342:252.
10. Rakela J, Lange SM, Ludwig J et al. Fulminant hepatitis: Mayo Clinic experience with 34 cases. Mayo Clinic Proc 1985; 60:289-292.
11. Castells A, Salmeron JM, Navasa M et al. Liver transplantation for acute liver failure: Analysis of applicability. Gastroenterology 1993; 105:532-538.
12. Gazzard BG, Portmann B, Murray-Lyon IM et al. Causes of death in fulminant hepatic failure and relationship to quantitative histologic assessments of parenchymal damage. Q J Med 1975; 44:615-626.
13. O'Grady JG, Alexander GJM, Hayllar KM et al. Early indicators of prognosis in fulminant hepatic failure. Gastroenterology 1989; 97:439-445.
14. Acute hepatic failure study group. Etiology and prognosis in fulminant hepatitis (Abstr). Gastroenterology 1979; 77:A33.

15. Bernuau J, Goudreau A, Poynard T. Hepatitis A virus infection: An uncommon cause of fulminant hepatitis with a high survival rate (Abstr). Hepatology 1983; 3:821.
16. Gimson AES, White YS, Eddleston ALWF et al. Clinical and prognostic differences in fulminant hepatitis type A, B and non-A, non-B. Gut 1983; 24:1194-1198.
17. Joshi YK, Gandhi BM, Tandon BN. Spectrum of hepatitis A virus infection in India (Abstr). Hepatology 1983; 3:1060.
18. Iwatsuki S, Steiber AC, Marsh JW et al. Liver transplantation of fulminant hepatic failure. Transplantation 1989; 21:2431-2434.
19. Sjogren MH, Tanno H, Fay O et al. Hepatitis A virus in stool during clinical relapse. Ann Intern Med 1987; 106:221-226.
20. Ritt DJ, Whelan G, Werner DJ et al. Acute hepatic necrosis with stupor or coma: An analysis of 31 patients. Medicine 1969; 48:151-172.
21. Redeker AG. Viral hepatitis. Am J Med Sci 1975; 270:9-16.
22. Lettau LA, McCarthy JG, Smith MH et al. Outbreak of severe hepatitis due to delta and hepatitis B virus in parenteral drug abusers and their contacts. N Engl J Med. 1987; 317:1256-1262.
23. Novick DM, Farci P, Croxson TS et al. Hepatitis D virus and HIV antibodies in parenteral drug abusers who are hepatitis B surface antigen positive. J Infect Dis 1988; 158:795-803.
24. Gimson AES, Tedder RS, White YS et al. Serological markers in fulminant hepatitis B. Gut 1983; 24:615-622.
25. Brechot C, Bernuau J, Thiers V et al. Multiplication of hepatitis B in fulminant hepatitis B. Br Med J 1984; 288:270-271.
26. Jones EA, Schafer DF. Fulminant hepatic failure. In: Zakim D, Boyer TD, eds. Hepatology: A textbook of liver disease. 2nd ed. Philadelphia: W.B.Saunders Company, 1990: 460-492.
27. Woolf IL, El Sheikh N, Cullens H et al. Enhanced hepatitis B surface antigen production in pathogenesis of fulminant viral hepatitis type B. Br Med J 1976; 2:669-671.
28. Samuel D, Bismuth A, Mathieu D et al. Passive immunoprophylaxis after liver transplantation in hepatitis B surface antigen-positive patients. Lancet 1991; 337:813-815.
29. Davis GL, Hoofnagle JH. Reactivation of chronic type B hepatitis presenting as acute viral hepatitis. Ann Intern Med 1985; 102: 762-767.
30. Levy P, Marcellin P, Martinot-Peignoux M et al. Clinical course of spontaneous reactivation of hepatitis B virus in patients with chronic hepatitis B virus. Hepatology 1990; 12:570-574.
31. Kotwal GJ, Baroudy BM, Kuramoto IK et al. Detection of acute hepatitis C infection by ELISA using a synthetic peptide comprising a structural epitope. Proc Natl Acad Sci 1992; 89:4486-4489.
32. Polito AJ, DiNello RK, Quan S et al. New generation RIBA hepatitis C strip immunoblot assays. Annal Biol 1992; 50:329-336.
33. Bukh J, Purcell RH, Miller RH. Importance of primer selection for the detection of hepatitis C virus with the PCR assay. Proc Natl Acad Sci 1992; 89:187-191.
34. Wright TL, Hsu H, Donegan E et al. Hepatitis C virus not found in fulminant non-A, non-B hepatitis. Ann Intern Med 1991; 115:111-112.
35. Feray C, Gigou M, Samuel D et al. Hepatitis C virus RNA and hepatitis B virus DNA in serum and liver of patients with fulminant hepatitis. Gastroenterology 1993; 104:549-555.
36. Liang TJ, Jeffers L, Reddy RK et al. Fulminant or subfulminant non-A, non-B hepatitis: The role of hepatitis C and E viruses. Gastroenterology 1993; 104:556-562.
37. Sallie R, Tibbs C, Silva AE et al. Detection of hepatitis E but not C in sera of patients with fulminant non-A, non-B hepatitis. Hepatology 1991; 14:68A. (abstract).
38. Yanagi M, Kaneko S, Unoura M et al. Hepatitis C virus in fulminant liver failure. N Engl J Med 1991; 324:1895-1896. (abstract).
39. Yoshiba M, Sekiyama K, Sugata F et al. Diagnosis of type C fulminant hepatitis by detection of antibodies to the putative core proteins of hepatitis C virus. Gastroenterol Japonicum 1991; 26:234.
40. Yoshiba M, Sekiyama K, Sugata F et al. Persistence of hepatitis C replication in non-A, non-B fulminant viral hepatitis. Gastroenterol Japonicum 1991; 26:235.

41. Villamil FG, Yu C-H, Hu K-Q et al. Detection of hepatitis C virus (HCV) by RNA polymerase chain reaction (PCR) in fulminant hepatic failure (FHF). Gastroenterology 1993; 104:1016A (abstract).
42. Woodfield DG. Acute post-transfusion fulminant hepatitis C virus infection: A case report. Gastroenterol Japonicum 1991; 26(Suppl 3):221-223.
43. Wright TL. Etiology of fulminant hepatic failure: Is another virus involved? Gastroenterology 1993; 104:640-643.
44. Mas A, Buti M, Esteban R et al. Hepatitis B virus and hepatitis D virus replication in hepatitis B surface antigen-positive fulminant failure. Hepatology 1990; 11:1062-1065.
45. Rizzetto M. The delta agent. Hepatology 1983; 3:729-737.
46. Saracco G, Macagno S, Rosina F et al. Serological markers with fulminant hepatitis in persons positive for hepatitis B surface antigen. Ann Intern Med 1988; 108:380-383.
47. Govindarajan S, Chin KP, Redeker AG et al. Fulminant B viral hepatitis: Role of delta agent. Gastroenterology 1984; 86:1417-1420.
48. Smedile A, Farci P, Verme G et al. Influence of delta infection on severity of hepatitis B. Lancet 1982; 2:945-947.
49. Purcell RH, Rizzett, M, Gerin JL. Hepatitis D infection of the liver. Semin Liver Dis 1984; 4:340-346.
50. Colombo M, Camberi R, Rumi MG et al. Long-term delta superinfection in hepatitis B surface antigen carriers and its relationship to the cause of chronic hepatitis. Gastroenterology 1983; 85:235-239.
51. Buti M, Esteban R, Espanol MT et al. Influence of HIV infectivity on cell-mediated immunity in chronic D hepatitis. J Infect Dis 1991; 163:1351-1353.
52. Lichtenstein DR, Makadon HJ, Chopra S. Fulminant hepatitis B and D virus coinfection in AIDS. Am J Gastroenterol 1992; 87:1643-1647.
53. Krawczynski K. Hepatitis E. Hepatology 1993; 17:932-941.
54. Centers for Disease Control. Hepatitis E among US travellers, 1989-1992. MMWR 1993; 42:1-4.
55. Zhuang H, Cao XY, Lin CB et al. Enterically transmitted non-A, non-B hepatitis is China. In: Shikata T, Purcell RH, Uchida T, eds. Viral hepatitis C,D,E. Amsterdam: Elsevier Science Publishers, 1991:227-285.
56. Centers for Disease Control. Enterically transmitted non-A, non-B hepatitis: East Africa. MMWR 1987; 36:241-244.
57. Centers for Disease Control. Enterically transmitted non-A, non-B hepatitis: Mexico. MMWR 1987; 36:597-602.
58. Valasquez O, Stetler HC, Avila C et al. Epidemic transmission of enterically transmitted non-A, non-B hepatitis in Mexico, 1986-1987. JAMA 1990; 263:3281-3285.
59. Wong DC, Purcell RH, Sreenivasan MA et al. Epidemic and endemic hepatitis in India: Evidence for non-A, non-B virus etiology. Lancet 1980; 2:876-879.
60. De Cock KM, Bradley DW, Sandford NL et al. Epidemic non-A, non-B hepatitis in patients from Pakistan. Ann Intern Med 1987; 106:227-230.
61. Arankalle VA, Ticehurst J, Sreenivasan MA et al. Aetiological association of a virus-like particle with enterically transmitted non-A, non-B hepatitis. Lancet 1988; 1:550-554.
62. Hyams KC, Purdy MA, Kaur M et al. Acute sporadic hepatitis E in Sudanese children: Analysis based on a new Western blot assay. J Infect Dis 1992; 165:1001-1005.
63. McCaustland KA, Bi S, Purdy MA et al. Application of two RNA extraction methods prior to amplification of hepatitis E virus nucleic acid by PCR. J Virol Methods 1991; 35:331-342.
64. Khuroo MS, Duermeyer W, Zargar SA et al. Acute sporadic non-A, non-B hepatitis in India. Am J Epidem 1983; 118:360-364.
65. Tandon BN, Gandhi BM, Joshi YK et al. Hepatitis virus non-A, non-B: The cause of a major public health problem in India. Bull WHO 1985; 63:931-934.
66. Khuroo MS, Teli MR, Skidmore S et al. Incidence and severity of viral hepatitis in pregnancy. Amer J Med 1981; 70:252-255.
67. Wainwright J, Schonland MM. Toxic hepatitis in black patients in Natal. S Afr Med J 1977; 51:571-573.

68. Katz M, Saibil FO. Herbal hepatitis: Subacute hepatic necrosis secondary to chaparral leaf. J Clin Gastroenterol 1990; 12:203-206.
69. Rose AG, Becker WB. Disseminated herpes virus hominis (herpes simplex) infection: A retrospective diagnosis for light and electron microscopy of paraffin wax-imbedded tissues. J Clin Path 1972; 25:79-87.
70. Klein NA, Mabie WC, Shaver DC et al. Herpes simplex virus hepatitis in pregnancy. Gastroenterology 1991; 100:239-244.
71. Elliot WC, Houghton DC, Bryant RE et al. Herpes simplex type 1 hepatitis in renal transplantation. Arch Int Med 1980; 140:1656-1660.
72. Shusterman NH, Frauenhoffer C, Kinsey MD. Fatal massive hepatic necrosis in cytomegalovirus mononucleosis. Ann Intern Med 1978; 88:810-812.
73. Markin RS, Linder J, Zuerlein K et al. Hepatitis in infectious mononucleosis. Gastroenterology 1987; 93:1210-1217.
74. Purtilo DT, Whit R, Filipovich A et al. Fulminant liver failure induced by adenovirus after bone marrow transplant (letter). N Engl J Med 1985; 312:1707-1708.
75. Phillips MJ, Blendis LM, Poucell S et al. Syncytial giant-cell hepatitis. N Engl J Med 1991; 324:455-460.
76. Zimmerman HJ. Hepatotoxicity: Adverse effects of drugs and other chemicals on the liver. New York: Appleton-Century-Crofts, 1978.
77. Farber E, Fisher MM. Toxic injury of the liver: Parts A and B. New York: Marcel Dekker Inc., 1980
78. Kaplowitz N, Aw TY, Simon FR et al. Drug-induced hepatotoxicity. UCLA conference. Ann Intern Med 1986; 104:826-839.
79. Prescott L, Illingsworth R, Critchley J et al. Intravenous N-acetylcysteine: Treatment of choice for paracetamol poisoning. Br Med J 1979; 2:1097-1100.
80. Harrison PM, Keays R, Bray GP et al. The effect of chronic alcohol intake on prognosis and outcome in paracetamol overdose. Human Experiment Toxicol 1991; 10:435-438.
81. Seeff LB, Cuccherini BA, Zimmerman HJ et al. Acetominophen hepatoxicity in alcoholics. Ann Intern Med 1986; 104:399-404.
82. Lauterberg BH, Velez ME. Glutathione deficiency in alcoholics: Risk factor for paracetamol hepatotoxicity. Gut 1988; 29:1153-1157.
83. Pessayre D, Bentata M, Degott C et al. Isoniazid-Rifampin fulminant hepatitis: A possible consequence of the enhancement of isoniazid hepatotoxicity by enzyme induction. Gastroenterology 1977; 72:284-289.
84. Mitchell JR, Zimmerman HJ, Ishak KG et al. Isoniazid liver injury: Clinical spectrum, pathology and probable pathogenesis. Ann Intern Med 1976; 84:181-192.
85. Dickinson DS, Bailey WC, Huschowitz BI et al. Risk factors for isoniazid (INH)-induced liver dysfunction. J Clin Gastroenterol 1981; 3:271-279.
86. Centers for Disease Control. Severe INH-associated hepatitis-New York, 1991-1993. MMWR 1993; 42:545-547.
87. Pessayre D, De Saint Louvert P, Degott C et al. Iproclozide fulminant hepatitis: Possible role of enzyme induction. Gastroenterology 1978; 75:492-496.
88. Danan G, Bernuau J, Moullot X et al. Amitriptylline-induced fulminant hepatitis. Digestion 1984; 30:179-184.
89. Rabinovitz G, Van Thiel DH. Hepatotoxicity of non-steroidal, anti-inflammatory drugs. Am J Gastroenterol 1983; 1696-1704.
90. Danan G, Trunet P, Bernuau J et al. Pirprofen-induced fulminant hepatitis. Gastroenterology 1985; 89:210-213.
91. Bravo JF, Jacobsen MP, Mertens BF. Fatty liver and pleural effusion with ibuprofen therapy. Ann Intern Med 1977; 87:200-201.
92. Kelsey WM, Scharjj M. Fatal hepatitis probably due to indomethacin. JAMA 1967; 199:586-587.
93. Zimmerman HJ. Update of hepatotoxicity due to classes of drugs in common clinical use: Non-steroidal drugs, anti-inflammatory drugs, antibiotics, antihypertensives, and cardiac and psychotropic agents. Sem Liver Dis 1990; 10:322-338.

94. Carney FMT, Van Dyke RA. Halothane hepatitis: A critical review. Anesth Analg 1972; 51:135-160.
95. Lewis JH, Zimmerman HJ, Ishak KG et al. Enflurane hepatotoxicity: A clinicopathological study of 24 cases. Ann Intern Med 1983; 98:984-992.
96. Faulstich H. New aspects of amanita poisoning. Klin Wochenschr 1979; 57:1143-1152.
97. Floersheim GL. Treatment of human amatoxin mushroom poisoning: Myths and advances in therapy. Med Toxicol 1987; 2:1-9.
98. Ruprah M, Mant TGK, Flanagan RJ. Acute carbon tetrachloride poisoning in 19 patients: Implications for diagnosis and treatment. Lancet 1985; 1:1027-1029.
99. Harrison R, Letz G, Pasternak B et al. Fulminant hepatic failure due to 2-nitropropane. Ann Intern Med 1987; 107:466-468.
100. Ngindu A, Johnson BK, Kenya PR et al. Outbreak of acute hepatitis caused by aflatoxin poisoning in Kenya. Lancet 1982; 1:1346-1348.
101. Roche-Sicot J, Benhamou J-P. Acute intravascular hemolysis and acute liver failure associated as a first manifestation of Wilson disease. Ann Intern Med 1997; 86:301-303.
102. Rector Jr WG, Ushida T, Kanel GC et al. Fulminant hepatitis and renal failure complicating Wilson's disease. Liver 1984; 4:341-347.
103. McCullough AJ, Fleming CR, Thistle JL et al. Diagnosis of Wilson's disease presenting as fulminant hepatic failure. Gastroenterology 1983; 84:161-167.
104. Peleman RP, Galavar JS, Van Thiel DH et al. Orthotopic liver transplant for acute and subacute hepatic failure in adults. Hepatology 1987; 7:484-489.
105. Stremmel W, Meyerose K-W, Niederau C et al. Wilson disease: Clinical prevention, treatment and survival. Ann Intern Med 1991; 115:720-726.
106. Gazzard BG, Portmann B, Murray-Lyon IM. Causes of death in fulminant hepatic failure and relationship to quantitative histological assessments of parenchymal damage. Q J Med 1975; 44:615-622.
107. Kanelis CT, Peters JL, Edmonson HA. Fatty liver of pregnancy and its relationship to tetracycline therapy. Amer J Med 1965; 38:359-364.
108. Pockros P, Peters RL, Reynolds TB. Idiopathic fatty liver of pregnancy: Finding in ten cases. Medicine (Baltimore) 1984; 63:1-11.
109. Bernuau J, Degott C, Nouel O et al. Non-fatal acute fatty liver of pregnancy. Gut 1983; 24:340-344.
110. Amon E, Allen SR, Petrie RH et al. Acute fatty liver of pregnancy associated with preeclampsia: Management of hepatic failure with postpartum liver transplantation. Am J Perinatol 1991; 8:278-279.
111. Croker JFS. Reye's syndrome. Sem Liver Dis 1982; 2:340-352.
112. Heubi JE, Partin JC, Partin JS et al. Reye's syndrome: Current concepts. Hepatology 1987; 7:155-164.
113. Arcidi Jr JM, Moore GW, Hutchins GM. Hepatic morphology in cardiac dysfunction: A clinicopathologic study of 1000 subjects at autopsy. Am J Pathol 1981; 104:159-166.
114. Nouel O, Henrion J, Bernuau J et al. Fulminant hepatic failure due to transient circulatory failure in patients with chronic heart disease. Dig Dis Sci 1980; 25:49-52.
115. Larner AJ, Dyer RG, Burke DA. Fulminant hepatic failure complicating exertional heatstroke with rhabdomyolosis. Eur J Gastroenterol Hepatology 1993; 5:55-58.
116. Hassanein T, Razack A, Gavaler JS et al. Heatstroke: Its clinical and pathological presentation, with particular emphasis to the liver. Am J Gastroenterol 1992; 87:1382-1389.
117. Sandle GI, Layton M, Record CO et al. Fulminant hepatic failure due to Budd-Chiari syndrome. Lancet 1980; 1:1199.
118. Mitchell JC, Bortnott JK, Kaufman S et al. Budd-Chiari syndrome: Etiology, diagnosis and management. Medicine 1982; 61:199-218.
119. Powell-Jackson PR, Ede RJ, Williams R. Budd-Chiari syndrome presenting as fulminant hepatic failure. Gut 1986; 27:1101-1105.
120. Shaked A, Goldstein RM, Klintmalm GB et al. Portosystemic shunt vs orthotopic liver transplant for Budd-Chiari syndrome. Surg Gynecol Obstet 1992; 174:453-459.

121. Zafrani ES, Pinaudeau Y, Dhumeaux D. Drug-induced vascular lesions of the liver. Arch Int Med 1983; 143:495-502.
122. Woods WG, Dekner LP, Nesbit ME. Fatal veno-occlusive disease of the liver following high dose chemotherapy, irradiation and bone marrow treatment. Amer J Med 1980; 68:285-290.
123. Griner PF, Elbadawi A, Packman CH. Veno-occlusive disease of the liver after chemotherapy of acute leukemia: Report of 2 cases. Ann Intern Med 1976; 85:578-582.
124. Bernuau J, Valli C, Boyer N. Fulminant hepatic failure due to autoimmune hepatitis (Abstr). J Hepatology 1987; 7:484-489.
125. Homberg J-C, Abuaf N, Bernard O et al. Chronic active hepatitis with antiliver/kidney microsomal antibody type 1: A second type of autoimmune hepatitis. Hepatology 1987; 7:1333-1339.
126. Eras P, Sherlock P. Hepatic coma secondary to metastatic liver disease. Ann Intern Med 1971; 14:581-583.
127. Harrison HB, Middleton III HM, Crosley JH et al. Fulminant hepatic failure: An unusual presentation of metastatic liver disease. Gastroenterology 1981; 80:820-825.
128. Zafrani ES, Leclercq B, Vernant JP et al. Massive blastic infiltration of the liver: A cause of fulminant hepatic failure. Hepatology 1983; 3:428-432.
129. Braude S, Gimson AES, Portmann B et al. Fulminant hepatic failure in non-Hodgkins lymphoma. Postgrad Med J 1982; 58:301-302.
130. Hsu SM, Xie SM, Hsu PL et al. Interleukin-6, but not interleukin-4 is expressed by Reed-Sternberg cells in Hodgkin's disease with or without histologic features of Castleman's disease. Am J Pathol 1992; 141:129-138.
131. Penn I. Hepatic transplantation for primary and metastatic cancer of the liver. Surgery 1991; 110:726-734.
132. Stein AM, Fawag K, Tabrizi A. Multifocal malignant hemangioendothelioma presenting as acute hepatitis: A clinicopathological study. Am J Gastroenterol 1977; 67:370-374.
133. Bismuth H, Houssin D, Mazmanian G. Post-operative liver insufficiency: Prevention and management. World J Surg 1985; 7:505-510.
134. Yamamaka N, Okamoto E, Kuwata K et al. A multiple regression equation for prediction of post hepatectomy liver failure. Ann Surg 1984; 200:653-663.
135. Takenaka K, Kanematsu T, Fukuzawa K et al. Can hepatic failure after surgery for hepatocellular carcinoma in cirrhotic patients be prevented? World J Surg 1990; 14:123-127.
136. Starzl TE, Demetris AJ, Van Thiel DH. Liver transplantation (First of two parts). N Engl J Med 1989; 321:1014-1022.
137. Alagille D, Odievre M. Liver and Biliary Tract Disease in Children. New York: John Wiley and Sons, 1979.
138. Psacharopoulos HT, Mowat AP, Davies M et al. Fulminant hepatic failure in childhood. Arch Dis Child 1980; 55:252-258.

CHAPTER 3

Acute Hepatic Encephalopathy: Pathophysiology and Diagnosis

Steven D. Colquhoun, Caroline A. Connelly

Hepatic encephalopathy is a neuropsychiatric syndrome that refers to potentially reversible derangement of thought and behavior which accompany liver disease.[1] Although there may be significant overlap in symptoms and etiology, both liver disease and hepatic encephalopathy can exist in either acute or chronic forms. Although chronic hepatic encephalopathy is more prevalent, the neurologic changes accompanying acute liver failure are more dramatic and may develop rapidly. Clinically, the management of these two entities can be quite distinct. Fulminant hepatic failure (FHF) is a true medical emergency and the onset of encephalopathy is both a defining symptom and a major prognostic indicator.[2] Without appropriate and timely intervention, the underlying neuropathology of this condition can lead directly to the patient's death. In the past decade, survival of FHF patients improved significantly due, in part, to enhanced ability to manage the cerebral manifestations of the disease.[2] This section will focus on the prevailing theories regarding the pathophysiology of the syndrome and will discuss the details regarding diagnosis and monitoring methods currently available.

Pathophysiology

Introduction

The association between liver disease and altered consciousness has been recognized for a millennium; moreover, serious research in this area has been under way for nearly a century. Nevertheless, there is still no complete understanding of what causes hepatic encephalopathy. Recent advances in technology have extended our understanding of both liver and brain physiology and function. Clinical and experimental findings now cover the spectrum from that of gross behavioral and physiologic changes to those at the biochemical and molecular levels. Unfortunately, cause and effect are difficult to discern when looking at functions as complex as those of the brain. Distinguishing primary from secondary changes among the astounding number of intricate and interrelated observations has proved to be an extraordinarily difficult task. Thus far, no single mechanism or unifying theory has arisen to explain the myriad of experimental and clinical findings that have been made. With the acute failure of synthetic, excretory and metabolic liver function, it would seem that a multitude of factors could potentially influence cerebral function. In fact, it now appears that the etiology of this syndrome is multifactorial and that a number of complex changes take place that manifest collectively as hepatic encephalopathy.[1]

Support of the Acutely Failing Liver, 2nd ed., edited by Achilles A. Demetriou and Frederick D. Watanabe.

Hepato-cerebral dysfunction is generally believed to occur when the brain is exposed to the presence of abnormal or toxic substances, rather than the absence of other factors that might be normally present or protective. It seems intuitive that the failure of hepatic clearance might lead to an accumulation of cerebral toxins. The most discussed candidates include substances such as ammonia and manganese, which in abnormal concentrations can have major global effects.[1] The influence of "endogenous" benzodiazepines on g-amino butyric acid (GABA)-ergic neurotransmission has also attracted attention. Actually, several neurotransmitters and neuromodulators are altered in acute hepatic encephalopathy, including GABA, glutamate, and nitric oxide.[1] The neuroglia, specifically astrocytes, which surround and support neuronal function, appear to be the primary cellular targets of pathologic change. Although many observations have been made, primarily in the setting of chronic encephalopathy, these same processes may also apply in the acute syndrome. Since the clinical manifestations of acute and chronic encephalopathy can be quite different, the likelihood of a common underlying pathophysiology remains, however, a matter of some debate. The mechanisms for each could be entirely distinct, have common areas of overlap, or they could be identical, with only the rate or magnitude of the pathologic change(s) accounting for the clinically observed differences.[3] Figure 3.1 summarizes the major factors contributing to hepatic encephalopathy.

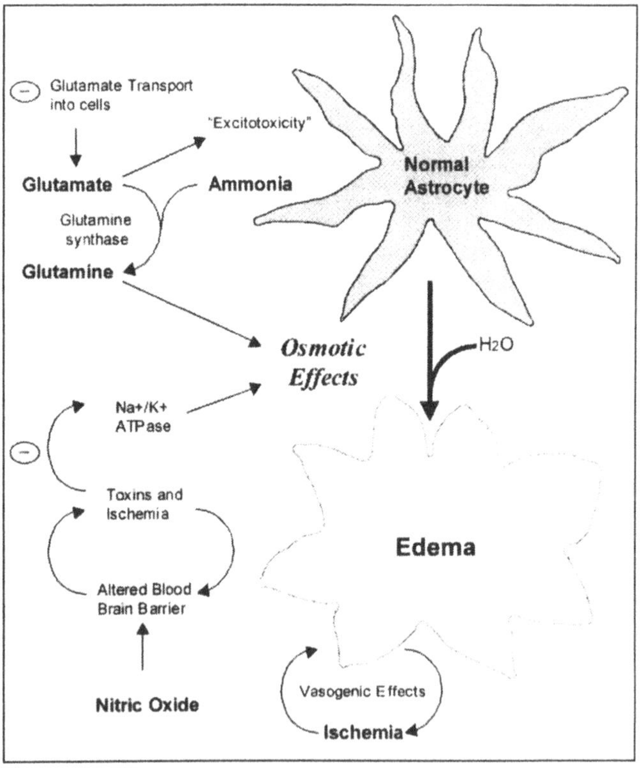

Fig. 3.1. Proposed factors contributing to hepatic encephalopathy.

Ammonia

Of all the theoretic changes thought to contribute to encephalopathy, the most pervasive and long-standing is that of ammonia. Elevated levels of nitrogenous compounds were first associated with mental status changes in early experiments utilizing the Eck fistula (portocaval anastomosis) in dogs.[4] In fact, the surgical bypass of hepatic detoxification results in elevated peripheral blood ammonia levels. Ammonia levels are frequently, but not consistently, elevated in humans with hepatic encephalopathy. Furthermore, the blood-brain barrier has an increased permeability for ammonia in patients with liver disease.[5,6] PET scan studies using $^{13}NH_3$ have shown an increased metabolism of ammonia in encephalopathic patients.[6] The presence of ammonia in the brain has extensive theoretic consequences, as will be discussed below.[1,7] The enzymatic machinery for the urea cycle is absent in the brain, which instead depends on glutamine synthesis. Ammonia has direct and indirect negative effects on both inhibitory and excitatory neurotransmission.[7,8] For example, ammonium ions impair inhibitory interactions in spontaneously active regions of the brain by inactivating the chloride extrusion necessary for inhibitory post-synaptic potentials to hyperpolarize the cell membrane. Ammonia also impairs excitatory glutamatergic neurotransmission and may play a role in brain edema formation.[5] Although the exact contribution of ammonia to acute encephalopathy remains unclear, it continues to be of major interest in the pathophysiology of this disorder.[7]

Cerebral Edema and Related Effects

General Considerations

Regardless of the underlying differences between acute and chronic encephalopathy, it is clear that alterations of blood flow and edema formation play a prominent role in the acute setting. Cerebral edema and intracranial hypertension occur in up to 80% of patients with acute liver failure.[9] The two most immediate causes of death in this group include herniation from cerebral edema and sepsis. Cerebral edema is also of major concern perioperatively in patients who are candidates for orthotopic liver transplantation, and such patients may have substantial residual neurologic deficits after recovery.[10]

A short interval between the onset of clinical jaundice and the first changes in mentation portends the occurrence of brain edema. It is surmised that, if this interval is otherwise sufficiently prolonged, compensatory osmotic changes take place and edema formation is averted. Indeed, clinical experience shows that cerebral edema is much less likely to occur in the subfulminant condition, and generally it only rarely occurs in patients with chronic hepatic encephalopathy. It is implied, but not confirmed, that the pathology of acute hepatic encephalopathy is shared with that of cerebral edema formation, although some have postulated parallel processes.[11] In any case, as a patient's condition deteriorates, the treatment of hepatic encephalopathy becomes exclusively a matter of managing brain swelling and intracranial pressure. As shall be discussed, edema per se can affect neurotransmission through alterations in extracellular electrolyte composition. Astrocytes play a key role in brain volume regulation, and injury or compromise of their function may further amplify the process of edema formation.

Vasogenic and Cytotoxic Mechanisms

There are two principal theories regarding the final common path leading to cerebral edema. The "vasogenic" theory postulates primary disruption of the blood-brain barrier with subsequent leakage of plasma into the extracellular space; the "cytotoxic" hypothesis postulates that a primary alteration of cellular osmoregulation causes direct intracellular edema.[3] In acute hepatic encephalopathy, it seems most likely that a cytotoxic mechanism predominates, although both may have a role during different phases of edema formation.[1,5] Regardless of the exact processes leading to cerebral edema, fundamental underlying events must occur and a number of possibilities have been proposed.

Cytokines and Inflammatory Factors

The release of toxic substances by necrotic liver cells is axiomatic and is strongly suspected to play some role in edema formation. This is supported by the clinical observation that patients with acute liver failure may respond to total hepatectomy with an immediate and significant lowering of ICP with improved overall hemodynamics.[12] Candidate substances include the inflammatory cytokines, such as IL-1, IL-6 or TNF-α.[2] The latter has, in fact, been shown to alter the permeability of the blood-brain barrier.[13] It has also been suggested that various elements of the arachidonic acid cascade may also be important contributing factors. Unfortunately, in experimental anhepatic models, animals may still proceed to develop cerebral edema, suggesting a more complex etiology for edema formation.

Cerebral Blood Flow

Primary alterations in cerebral blood flow (CBF) have been postulated to play a role in the genesis of cerebral edema and elevation of intracranial pressure (ICP). The determinants of ICP are described by the Monroe-Kellie doctrine: Within the fixed space of the cranium, an increase in any of the three compartments of brain, cerebrospinal fluid (CSF), or blood will cause an increase in pressure. Under normal conditions, CBF is under autoregulatory control; meaning that blood flow remains relatively constant over a wide range of mean arterial pressure (MAP). Whether CBF changes are primary or secondary to cerebral edema is not clear. Cerebral blood flow studies in both animals and humans have provided conflicting information.[14] In experimental studies of acute liver failure in rats, it has been shown that a linear decrease in cerebral blood flow precedes a rise in ICP.[11] This has been presumed to occur via auto-regulatory changes responding to diminished metabolic activity.[11] It has also been speculated that primary loss of auto-regulation may occur in acute liver failure, resulting in increased blood flow to the brain compared to the cerebral metabolic demands.[15] This phenomenon of cerebral hyperemia has been termed "cerebral luxury perfusion" and is felt to primarily contribute to cerebral edema formation.[16] Alternatively, loss of auto-regulation could result in cerebral hypoperfusion and hypoxemia, which could also contribute to edema.[17]

By whatever mechanism, even a small amount of edema may compromise tissue oxygenation through an increase in the diffusion distance from brain capillaries to cells. A vicious cycle of edema and ischemia may then occur. Neuronal function is likely affected by both alterations in extracellular electrolyte composition and ischemic compromise. Astrocytes also play a key role in brain volume regulation and, as we shall see, appear to play a large part in both edema formation and neuronal dysfunction.

Glutamine

Glutamine is an important organic osmolyte appearing to strongly participate in the generation of cerebral edema. In the neuroglia, glutamine is formed when ammonia binds to glutamate in a reaction catalyzed by glutamine synthetase. In acute liver failure, the presence of excessive ammonia drives this reaction toward accumulation of glutamine.[5] In the absence of compensatory changes, fluid shifts into glial cells and contributes to tissue edema. Indeed, cerebral edema and elevated ICP occur in other conditions of hyperammonemia such as Reye's syndrome and ornithine decarboxylase deficiency. In animal models, ammonia infusion alone can result in elevated ICP, and the inhibition of glutamine synthetase can provide protection from ammonia in both in vivo and in vitro models. Glutamine levels are increased in animal models of acute liver failure, in humans examined by MRI spectroscopy and in autopsy studies.[18,19] Interestingly, glutamine is also elevated in the brains of patients with chronic disease; however, such patients have correspondingly lower levels of other osmotic agents such as myoinositol, as seen by MR sepectroscopy.[5,20] Experimental evidence suggests that in the chronic setting, compensatory changes have been allowed to occur, thus avoiding edema formation.[21]

Glutamate

In addition to being a critical substrate necessary for the synthesis of glutamine, glutamate is also an important excitatory neurotransmitter in the central nervous system.[22] In acute liver failure, total brain glutamate content is reduced in both patients and animal models.[7,23] At the same time, the glutamate required for excitatory synaptic transmission, found in the extracellular spaces of the brain, is actually increased.[24-26] Excess glutamate in the extracellular space apparently results from impaired transport of glutamate from the extracellular to intracellular space and correlates well with blood ammonia concentrations.[24] Consistent with this theory, the GLAST and GLT-1 transporter proteins responsible for moving glutamate across the membrane into the cells, and the mRNA encoding for these proteins, are both decreased in comatose rats with acute liver failure.[27] Elevation of ammonia in the brain (1-3 mM) appears to signal the downregulation of these glutamate transporter proteins.

The N-methyl-D-aspartate (NMDA) and non-NMDA subclasses of excitatory amino acid receptors are involved in transmission of excitatory signals between neurons in the brain. The non-NMDA receptors are further subdivided into the kainate and AMPA subgroups. Excess glutamate in the synaptic cleft may impair neurotransmission by desensitizing AMPA receptors, which have decreased binding capacity in rat and rabbit models of acute liver failure.[25,28] The densities of non-NMDA binding sites are decreased in animal models of acute liver failure, but the NMDA binding sites are unchanged.[7,24,26,28] NMDA receptors may also mediate deleterious effects on brain function through other mechanisms. Overstimulation of NMDA receptors by excess extracellular glutamate may gate a neurotoxic excess entry of calcium ions into neurons.[3] NMDA receptor-mediated entry of calcium to the intracellular compartment can trigger production of nitric oxide, which may contribute to the increased permeability of the blood-brain barrier. In addition, glutamate infusions have experimentally been shown to cause cerebral edema, possibly through astrocyte AMPA activation.[7]

Nitric Oxide

Excess production of nitric oxide (NO) and oxidants derived from NO, such as peroxynitrite, may be involved in the pathogenesis of hepatic encephalopathy (Fig. 3.2). Nitric oxide and its metabolites may play a role in many of the systemic manifestations of acute liver failure, including those which are neurologic.[29] The metabolic products of NO have been shown to be elevated in the plasma of patients with acute liver failure.[29] Nitric oxide (NO) affects large regions by rapidly diffusing from one neuron to another to act directly on intracellular components. Because NO is permeable to membranes, it cannot be stored or concentrated by cells. Rather, NO activity is regulated through its synthesis by the enzyme nitric oxide synthase (NOS). NOS expression correlates locally with NO activity. Endothelial constitutive NOS (ecNOS) and nitrotyrosine, a metabolic endproduct indicative of excess nitric oxide production, are both regionally increased in the brains of portocaval-shunted rats (Table 3.1).[30] Increased expression of the mRNA encoding for neuronal NOS has also been observed in animal models of chronic encephalopathy.[31]

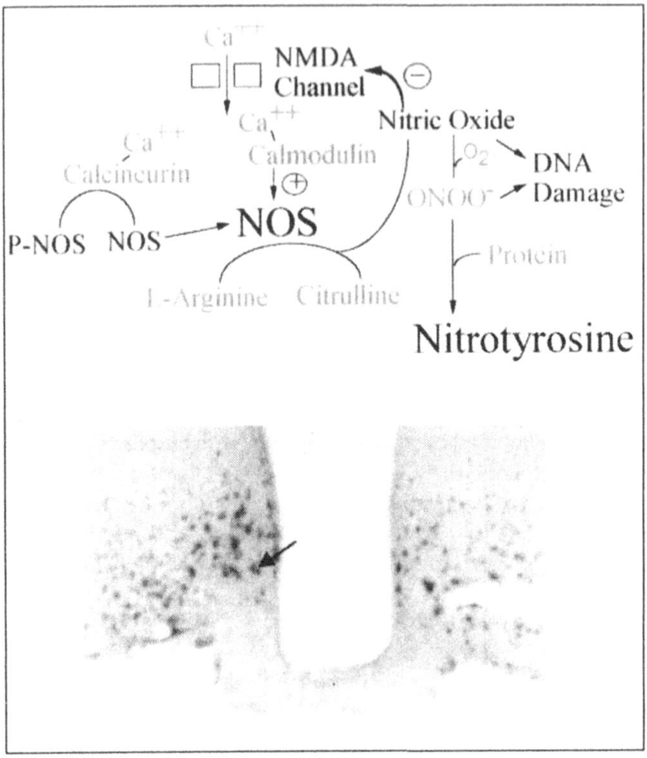

Fig. 3.2.

The permanent deposition of nitrotyrosine in glia and neurons of the hypothalamus and hippocampus may underlie impaired sleep patterns, mood and memory deficits associated with hepatic encephalopathy. Increased NO and nitrotyrosine levels in endothelial cells and glia lining blood vessels and ventricles of the brain may be associated with the observed increased permeability of the blood-brain barrier in this syndrome. Excess NO can directly inhibit NMDA receptor function of neurons and cause regional DNA damage in brain tissue. In the chronic setting, ammonia is the suspected trigger for the stimulation of ecNOS production in the brain. In acute liver failure. ammonia may also trigger the induction of NOS and a similar cascade of NO-mediated events.

Astrocytes

Brain edema in acute liver failure is due in large part to swelling of the star-shaped neuroglial astrocytes surrounding neurons and supporting their function.[1,32] Astrocytes are critically important for the maintenance of the extracellular microenvironment necessary to support the function of neurons. Initially described simply as "neural glue", the role of glia has expanded as it has become clear that these cells participate in the uptake and release of neurotransmitters, ions and neuromodulators which profoundly impact the efficacy of signal transmission in the brain. Astrocytes, rather than neurons, appear to be the major site of morphologic change in hepatic encephalopathy.

Astrocytes are involved in osmo-regulation as well as detoxification of substances entering the cerebral circulation, including ammonia, drugs and hormones. In fact, astrocytes

Table 3.1. Forebrain regions of portocaval shunted rats (10 months after surgery) which were immunohistochemically labeled with antibodies against endothelial constitutive nitric oxide synthase (ecNOS) or nitrotyrosine (NT).[30]

	ecNOS	NT
Hypothalamus	+	+
Piriform Cortex	+	+
Hippocampus	+	+
Supraoptic Nucleus	+	+
Amygdala	+	+
Circumventricular Organs	+	+
Lining of the Third and Lateral Ventricles	+	+
Blood Vessel Endothelium	+	0

(A + or 0 indicate presence or absence of increased expression of mRNA.)

possess the enzyme glutamine synthetase and are the only cells capable of metabolizing ammonia in the brain.[32] In addition, these cells participate in the scavenging of free radicals, agents of the inflammatory/immune response and they appear to significantly affect neuronal excitability and neurotransmission.[33] Astrocytes represent a key element of the "blood-brain barrier", which has repeatedly been shown to become more permeable in animal models of acute and chronic hepatic encephalopathy. Numerous pathologic studies have also described abnormal changes in astrocytes associated with encephalopathy. Histologic and biochemical changes in astrocytes are associated with chronic hepatic encephalopathy, including abnormal glutamate kinetics, altered benzodiazepine receptor expression, reduced enzyme function and neurotransmitter regulation.[1,5] Astrocytes are currently a primary focus in attempts to understand the pathophysiology of hepatic encephalopathy.[34,35]

Na^+/K^+ Pump

In animal models of acute liver failure, alterations in the neuronal active transport of sodium and potassium have also been observed. Circulating toxins or conditions of ischemia may inhibit Na^+K^+ ATPase. The subsequent accumulation of intracellular sodium could contribute to edema formation, giving rise to a cycle of ischemia and edema that may accelerate pathologic conditions.[36] A number of animal models and clinical observations support the idea that an element of ischemia may be associated with abnormal sodium ion transport. Once again, whether such observations represent primary or secondary pathologic changes remains to be determined.

Other Mechanisms

Benzodiazepines, GABA and Neurosteroids

Endogenous or exogenously ingested benzodiazepines may also play a role in the etiology of acute and chronic hepatic encephalopathy. In the past several years, dozens of basic science and clinical reports have appeared implicating these substances which enhance the inhibitory GABA-ergic tone. The gut flora has been postulated to contribute benzodiazepine ligand activity, which fits well with the general theory regarding the functional failure of hepatic clearance.[37] Indeed, animal models of acute liver failure have demonstrated increased benzodiazepine receptor agonist activity.[38] Clinical studies in patients with fulminant liver failure have documented an increase in benzodiazepine receptor ligands with enhanced GABA-ergic tone in all stages of encephalopathy.[29,39] Increased densities of brain peripheral-type benzodiazepine receptors, and elevated levels of neurosteroids, have also been reported in animal models of acute liver failure.[40,41] It is suggested that ammonia-induced activation of peripheral-type benzodiazepine receptors on astrocytes causes release of neurosteroids which enhance GABA-ergic neurotransmission. Recently, Norenberg et al demonstrated that neurosteroids impair the ability of cultured astrocytes to take up GABA in the acute setting.[33] Hence, neurosteroids may act to potentiate the inhibitory actions of GABA at the level of the receptor and also to lengthen the duration of time that the GABA ligand remains in the synaptic cleft. Ammonia also acts directly to potentiate GABA-ergic tone by enhancing the affinity of GABA for GABA-A receptors.[42]

Despite these seemingly compelling findings, no direct clinical relationship has been shown between either benzodiazepine ligands or their receptors and acute encephalopathy. There is no apparent correlation between ligand and clinical stage of encephalopathy, nor is ligand elevation a consistent finding among all patients.[39] Furthermore, clinical and animal studies in which a benzodiazepine receptor antagonist has been administered have not shown consistent effects.[43-45] The anecdotal success of flumazenil, the principal benzodiazepine antagonist, has been ascribed to the antagonism of benzodiazepines ingested by patients. Support for the concept of "endogenous" benzodiazepines arising from the intestine can be found in a number of studies, including those in dogs with congenital portocaval shunts, showing ligand activity in stool samples.[46] Whether there is a causal relationship and/or these represent associated phenomena, remains to be determined. Whatever the contribution of such substances to the pathophysiology of acute hepatic encephalopathy and cerebral edema, it appears likely that other mechanisms are involved.

Energy Metabolism

One early concept regarding the possible etiology of hepatic encephalopathy focused on a primary lack of cerebral energy as a result of hepatic metabolic failure.[5] A variety of techniques have been used to suggest that hypometabolism may be linked to this condition. Positron emission tomography (PET) and single-photon emission computed tomography (SPECT) scans have both demonstrated decreased cerebral glucose utilization in key regions of the brain.[1] However, it remains to be demonstrated whether such observations represent primary or secondary changes, or play any role in the acute setting. Other studies have looked at high energy phosphate bonds utilizing cerebral ^{31}P magnetic resonance spectroscopy and found no primary deficit in energy metabolism in such patients.[47] A primary lack of metabolic energy per se now seems less likely than changes leading to failure of neurotransmission and secondary alterations of energy utilization.[5]

Manganese and Toxic Metals

In the past, only postmortem methods were available to study clinical changes occurring in the brains of patients with acute or chronic hepatic encephalopathy.[1] Such studies have identified a number of alterations in such patients, including elevated concentrations of toxic metals in the brain. Such techniques, however, are often suspect due to the potential for many unrelated changes that can occur in the perimortem period. The introduction of magnetic resonance imaging (MRI) to this area has allowed observations of dynamic changes in the brains of living patients with these syndromes.[1] Manganese is ordinarily cleared by the liver, but remains in the circulation of patients with liver damage.[1] Enhanced T1-weighted MRI signals in the globus pallidus have been correlated with elevated brain manganese levels measured in patients dying in hepatic coma.[48] It is conceivable that depositions of manganese in pallidal regions may contribute to the extrapyramidal symptoms often noted in patients with liver disease. The use of chelating agents as a therapeutic option to reduce the neurologic symptoms in these patients has not been well studied.

Following galactosamine administration, the content of zinc has been shown to be elevated in the brains of rats.[49] However, in another study looking at autopsy specimens from nine patients with hepatic encephalopathy, zinc levels in the globus pallidus were normal, while manganese and copper concentrations were found to be elevated.[48]

Seizure Activity

It has been observed that some patients with acute liver failure may develop seizure activity which may markedly affect both cerebral blood flow and brain oxygen utilization.[1,29] Such activity may go unrecognized in the setting of mechanical ventilation, sedation and pharmacologic paralysis. It has been suggested that unrecognized seizure activity could contribute to the formation of cerebral edema.[50] Although it is certainly important to recognize and treat seizure activity, as a mechanism for edema formation this has been met with skepticism. Most neurologists can cite examples of patients with unrecognized and untreated status epilepticus who did not develop cerebral edema.

Diagnosis

General Considerations

In contrast to the insidious onset and relapsing nature of chronic hepatic encephalopathy, the clinical syndrome associated with acute liver failure progresses from mild alterations of mental status to coma within days, or even hours. Often a change in mental status can be the most overt clinical sign of acute liver failure.[2] The combination of significant coagulopathy and mental status changes is ominous. Altered mental status in a patient even suspected of having acute liver failure is an indication for admission to a critical care unit and should initiate contact with a transplant center.[51] The window of opportunity for transferring a patient while still relatively stable may rapidly close. If a patient is potentially a transplant candidate, then clearly it is in his/her best interest to be at a transplant center.

Neurologic Examination

The earliest changes of hepatic encephalopathy include only minor variations in mental status. Increased irritability, alterations in sleep/wake cycle and subtle personality changes and agitation may occur. The stages of hepatic coma are listed in Table 3.2. Neurologically, early clinical features include a generalized increase in muscle tone which may later progress to decerebrate posturing. A brisk deep tendon reflex may also be an early sign, with sustained clonus manifested at later stages. Psychomotor agitation is a common feature, leading to inappropriate sedative administration by inexperienced clinicians. In later stages, seizures can further complicate the clinical picture. Localizing signs, in general, are not part of hepatic encephalopathy and rather suggest another etiology or an event such as an intracranial hemorrhage. Spontaneous hyperventilation resulting in significant respiratory alkalosis is a clinical sign that can occur but is usually lost with intubation and ventilation. Once patients are moved to the intensive care unit, endotracheally intubated, mechanically ventilated and perhaps sedated or paralyzed, there are very few clinical neurologic signs remaining which can be used to follow clinical progression of the disease. However, pupillary response persists and may become the most important remaining clinical parameter: Slowing of the pupillary response correlates with progression of cerebral edema.

Table 3.2. Grading of encephalopathy

Stage 0	Normal
Stage 1	Sleep disturbances Personality Changes
Stage 2	Disorientation Lethargy
Stage 3	Stupor Arousable
Stage 4	Coma

Imaging Studies

There is no diagnostic role for brain imaging studies in the acute setting of hepatic encephalopathy. Computed tomography (CT) or magnetic resonance imaging (MRI) scans are indicated only when there are focal neurologic signs or concern regarding an event such as intracranial hemorrhage. In addition, many such patients are not clinically stable for transport to the radiology suite. It should be emphasized that the diagnosis of cerebral edema is based on clinical criteria alone, rather than the results of a particular imaging study. If intracranial hemorrhage is suspected, then a CT scan may be indicated to facilitate or terminate further treatment. When a CT scan is obtained in the context of elevated ICP, a spectrum of findings may be seen, ranging from ventricular collapse to little or no change.[29]

Laboratory Studies

Routine blood tests are usually carried out. The assessment of hepatic synthetic function is of primary interest, with the focus on the most responsive indicators, such as prothrombin time and INR. A prothrombin time greater than 20 seconds, an INR over 2 or a worsening trend are worrisome. Liver enzymes are usually followed as a reflection of ongoing hepatocyte necrosis. Absolute numbers may be astoundingly high, but have little consequence toward management; trends are much more telling. The combination of falling enzymes and rising INR is particularly ominous. The hematocrit is followed to rule out occult bleeding or hemolysis. The white blood cell count is also helpful to look for evidence of infection or signs of neutropenia. The prognostic criteria as developed at King's College are discussed elsewhere in this monograph.

There are few tests pertaining specifically to acute encephalopathy or cerebral edema. Despite the potential involvement of ammonia in the pathophysiology of this condition, serum levels correlate poorly with clinical progress and have no significance for either management or determination of prognosis. Diuresis may be important in managing cerebral edema and it may therefore be necessary to monitor serum osmolarity.

Other Diagnostic Measures

As discussed earlier, there have been conflicting reports regarding the significance of cerebral blood flow (CBF) alterations in acute liver failure; studies suggest both increased and decreased flows.[29] Jugular venous oxygen saturation was suggested as an important tool for the diagnosis of cerebral hyperemia (when values are elevated above 75%). Cerebral hyperperfusion has been documented immediately preceding brain death in patients with FHF.

Electroencephalographic (EEG) monitoring has been advocated by some in patients with acute liver failure to avoid unrecognized seizure activity.[29] However, the EEG per se shows no characteristic patterns diagnostic of hepatic encephalopathy, but rather only the diffuse changes of metabolic dysfunction. The EEG may be utilized as a confirmatory test when symptoms have progressed and brain death is suspected. Other confirmatory tests would include a nuclear medicine cerebral perfusion scan or angiography showing no cerebral perfusion.

Summary

Acute hepatic encephalopathy is a disorder representing the nexus of the two most complex organs of the body, and is clearly an integral aspect of acute liver failure. Its presence defines FHF and its progression reflects its prognosis. For the scientist, the pathophysiology of this syndrome is a fascinating and active area of research, bridging numerous independent disciplines in new and creative ways. For the clinician, acute hepatic encephalopathy represents a challenging aspect of the FHF syndrome, both from a diagnostic as well as a therapeutic perspective.

References

1. Butterworth RF. The neurobiology of hepatic encephalopathy. Semin Liver Dis 1996; 16:235-244.
2. Williams R. Classification, etiology and considerations of outcome in acute liver failure. Semin Liver Dis 1996; 16:343-345.
3. Blei AT. Pathogenesis of brain edema in fulminant hepatic failure. In: Progress in Liver Diseases, Boyer JL, Ockner RK, eds. Philadelphia, W.B. Saunders, 1995; 311-330.
4. Hahn M, Massen O, Nencki M, Pavlov J. Die ecksche fistel zwischen der unteren hohlvene und der Pfortader und ihre folgen fur den organismus. Arch Exper Pathol and Pharmakol 1893; 32:161.
5. Mousseau DD, Butterworth RF. Current theories on the pathogenesis of hepatic encephalopathy. Proc Soc Exper Biol and Med 1994; 206:329-344.
6. Lockwood AH, Yap EWH, Wong WH. Cerebral ammonia metabolism in patients with severe liver disease and minimal hepatic encephalopathy. J Cerebr Bl Flow Metabol 1991; 11:337-341.
7. Michalak A, Knecht K, Butterworth RF. Hepatic encephalopathy in acute liver failure: Role of the glutamate system. Adv Exper Med Biol 1997; 420:35-43.
8. Raabe WA. Neurophysiology of ammonia intoxication. In: Hepatic Encephalopathy; Pathophysiology and Treatment, Butterworth RF, Layrargues GP, eds. Clifton, New Jersey, Humana Press, 1989; 49.
9. Ede RJ, Gimson AE, Bihari D et al. Controlled hyperventilation in the prevention of cerebral edema in FHF. J Hepatol 1986; 2:43-51.
10. Prager MC, Washington DE, Lidofsky SD et al. Intracranial pressure monitoring during liver transplant without veno-venous bypass for FHF. Transpl Proc 1993; 25:1841.
11. Shah V, Webster S, Gottstein J et al. Reduction of cerebral perfusion precedes rise of intracranial pressure in rats with ischemic fulminant liver failure. Hepatology 1993; 17:1117-1122.
12. Ejlersen E, Larsen FS, Pott F et al. Hepatectomy corrects cerebral hyperperfusion in FHF. Transpl Proc 1994; 26:1794-1795.
13. Odeh M. Endotoxin and tumor necrosis factor-alpha in the pathogenesis of hepatic encephalopathy. J Clin Gastroenter 1994; 19:146-153.
14. Wendon JA, Harrison PM, Keays R et al. Cerebral blood flow and metabolism in fulminant liver failure. Hepatology 1994; 19:1407-1413.
15. Larsen FS, Knudsen GM, Paulson OB et al. Cerebral blood flow autoregulation is absent in rats with thioacetamde-induced hepatic failure. J Hepatol 1994; 21:491-495.

16. Larsen FS, Hansen BA. Cerebral luxury perfusion in fulminant hepatic failure. In: Advances in Hepatic Encephalopathy and Metabolism in Liver Disease, Record C, Al-Mardini H, eds. Newcastle Upon Tyne, Ipswich Book Company, 1997; 421.
17. Larsen FS, Ejlersen E, Hansen BA et al. Functional loss of cerebral blood flow autoregulation in patients with fulminant hepatic failure. J Hepatol 1995; 23:212-217.
18. McConnell JR, Antonson DL, Ong CS et al. Proton spectroscopy of brain glutamine in acute liver failure. Hepatology 1995; 22:69-74.
19. Record CO, Buxton B, Chase R et al. Plasma and brain amino acids in fulminant hepatic failure and their relationship to hepatic encephalopathy. Eur J Clin Invest 1976; 6:387-394.
20. Cordoba J, Gottstein J, Blei AT. Glutamine, myo-inositol, and organic brain osmolytes after portocaval anastomosis in the rat: Implications for ammonia-induced brain edema. Hepatology 1996; 24:919-923.
21. Cordoba J. Glutamine, myo-inositol, brain edema in acute liver failure [letter]. Hepatology 1996; 23:1291.
22. Mayer ML, Westbrook GL. The physiology of excitatory amino acids in the vertebrate central nervous system. Progr Neurobiol 1997; 28:197.
23. Swain M, Butterworth RF, Blei AT. Ammonia and related amino acids in the pathogenesis of brain edema in acute ischemic liver failure in rats. Hepatology 1992; 15:449-453.
24. Butterworth RF. Neuroactive amino acids in hepatic encephalopathy. Metabol Brain Dis 1996; 11:165-173.
25. Michalak A, Rose C, Butterworth J et al. Neuroactive amino acids and glutamate (NMDA) receptors in frontal cortex of rats with experimental acute liver failure. Hepatology 1996; 24:908-913.
26. de Knegt RJ, Schalm SW, van der Rijt CC et al. Extracellular brain glutamate during acute liver failure and during acute hyperammonemia simulating acute liver failure. J Hepatol 1994; 20:19-26.
27. Knecht K, Michalak A, Rose C et al. Decreased glutamate transporter (GLT-1) expression in frontal cortex of rats with acute liver failure. Neuroscience Letters 1997; 229:201-203.
28. Ferenci P. Brain dysfunction in FHF. J Hepatol 1994; 21:487-490.
29. Ellis A, Wendon J. Circulatory, respiratory, cerebral, and renal derangements in acute liver failure: Pathophysiology and management. Semin Liver Dis 1997; 16:379-388.
30. Connelly CA, Zardetto-Smith A, Colquhoun SD. Increased nitrotyrosine immunoreactivity in forebrain regions of portocaval-shunted rats. Society of Neuroscience Abstracts 1997; 23:1232.(Abstract)
31. Rao VL, Audet RM, Butterworth RF. Increased neuronal nitric oxide synthase expression in brain following portocaval anastomosis. Brain Res 1997; 765:169-172.
32. Blei AT. Brain edema—its relation to hepatic encephalopathy. In: Advances in Hepatic Encephalopathy and Metabolism in Liver Disease, Record C, Al-Mardini H, eds. Newcastle Upon Tyne, Ipswich Book Company, 1997; 259.
33. Norenberg MD, Itzhak Y, Bender AS. The peripheral benzodiazepine receptor and neurosteroids in hepatic encephalopathy. Adv Exper Med Biol 1997; 420:95-111.
34. Swain MS, Bergeron M, Audet R et al. Monitoring of neurotransmitter amino acids by means of an indwelling cisterna magna catheter. Hepatology 1992; 16:1028-1035.
35. Butterworth RF. Taurine in hepatic encephalopathy. Adv Exper Med Biol 1996; 403:601-606.
36. Cordoba J, Blei AT. Brain edema and hepatic encephalopathy. Semin Liver Dis 1996; 16:271-280.
37. Yurdaydin C, Walsh TJ, Engler HD et al. Gut bacteria provide precursors of benzodiazepine receptor ligands in a rat model of hepatic encephalopathy. Brain Res 1995; 679:42-48.
38. Jones EA, Basile AS, Skolnick P. Hepatic encephalopathy, GABA-ergic neurotransmission and benzodiazepine receptor ligands. Adv Exper Med Biol 1990; 272:121-134.
39. Basile AS, Harrison PM, Hughes RD et al. Relationship between plasma benzodiazepine receptor ligand concentrations and severity of hepatic encephalopathy. Hepatology 1994; 19:112-121.

40. Kadota Y, Inoue K, Tokunaga R et al. Induction of peripheral-type benzodiazepine receptors in mouse brain following thioacetamide-induced acute liver failure. Life Sci 1996; 58:953-959.
41. Itzhak Y, Roig-Cantisano A, Dombro RS et al. Acute liver failure and hyperammonemia increase peripheral-type benzodiazepine receptor binding and pregnenolone synthesis in mouse brain. Brain Res 1995; 705:345-348.
42. Takahashi K, Kameda H, Kataoka M et al. Ammonia potentiates GABAA response in dissociated rat cortical neurons. Neuroscience Letters 1993; 151:51-54.
43. Sutherland LR, Minuk GY. Ro 15-1788 and hepatic failure. [Letter] Ann Int Med 1988; 108:158.
44. Steindl P, Puspok A, Druml W et al. Beneficial effect of pharmacological modulation of the GABAA-benzodiazepine receptor on hepatic encephalopathy in the rat: Comparison with uremic encephalopathy. Hepatology 1991; 14:963-968.
45. Widler P, Fisch HU, Schoch P et al. Increased benzodiazepine-like activity is neither necessary nor sufficient to explain acute hepatic encephalopathy in the thioacetamide-treated rat. Hepatology 1993; 18:1459-1464.
46. Mullen KD, Gacad R, Kaminsky-Russ K et al. "Endogenous" benzodiazepines may originate from the intestine. In: Advances in Hepatic Encephalopathy and Metabolism in Liver Disease, Record C, Al-Mardini H, eds. Newcastle Upon Tyne, Ipswich Book Company, 1997; 239.
47. Morgan MY. Noninvasive neuroinvestigation in liver disease. Semin Liver Dis 1996; 16:293-314.
48. Layrargues GP, Spahr L, Fontaine L et al. T1-wieghted magnetic resonance (MR) signal hyperintensity in globus pallidus of cirrhotic patients results from manganese deposition. In: Advances in Hepatic Encephalopathy and Metabolism in Liver Disease, Record C, Al-Mardini H, eds. Newcastle Upon Tyne, Ipswich Book Company, 1997; 251.
49. Hu HL, Chen RD. Changes in free radicals, trace elements, and neurophysiological function in rats with liver damage induced by D-galactosamine. Biol Trace Elem Res 1992; 34:19-25.
50. Ellis AJ, Wendon JA, Williams R. Does subclinical seizure activity cause cerebral edema in patients with acute liver failure? International Symposium on Hepatocerebral Disorders: 1997; Val David, Quebec: 5.(Abstract)
51. Lee WM. Management of acute liver failure. Semin Liver Dis 1997; 16:369-378.

CHAPTER 4

Acute Hepatic Encephalopathy: Treatment

Olivier Detry, Jody E. Margulies, Nicholas Arkadopoulos, Achilles A. Demetriou

Intracranial hypertension secondary to brain edema is the most common cause of death in fulminant hepatic failure (FHF) patients, even when they are listed for urgent orthotopic liver transplantation (OLT).[1] Brain edema is a complication of FHF. In a 1944 report of 125 autopsies of FHF patients, Lucké did not find evidence of cerebral herniation.[2] Cerebral edema was noted in a few cases, but his finding could in part be attributed to brain ischemia. The first reports of brain edema and cerebral herniation as FHF complications were published in the 1970s, but wide recognition of this complication came only in the 1980s.[3-6] One possible explanation for the relatively recent recognition of this complication may be the advances in FHF patient care. Previously, FHF patients were dying from complications of early hepato-cellular insufficiency, mainly hemorrhage and sepsis.[2] Improvements in intensive care techniques have enhanced the survival of FHF patients. The longer course of the disease may result in development of brain edema, which may represent a later manifestation of FHF. In the current management of FHF patients, intracranial hypertension remains the greatest clinical challenge.

Standard Treatment

Our goal in the medical management of FHF patients with brain edema is to maintain the intracranial pressure (ICP) below 20 mmHg and the cerebral perfusion pressure (CPP) above 70 mmHg. Cerebral ischemia occurs if CPP is less than 40-50 mmHg and OLT is generally contraindicated if CPP remains below 40 mmHg for two hours.[1] To achieve this goal requires intense medical management.

FHF patients are admitted to a dedicated unit, and are monitored with peripheral arterial and pulmonary artery catheters after blood product infusion to partially correct their coagulopathy. FHF patients are treated with standard supportive measures to correct electrolyte, metabolic, respiratory and hemodynamic abnormalities. Euvolemia is maintained with regular measurements of pulmonary artery wedge pressure; hypovolemia may exacerbate hypotension and may reduce CPP and cerebral blood flow (CBF), resulting in brain ischemia. Systemic hypertension may also be deleterious by increasing the ICP.[5] Treating hypertension with β-blockers may be more useful than nitroprusside or calcium channel inhibitors, because of the absence of brain blood vessel dilatation. Appropriate patient positioning and nursing are also important in the care of FHF patients with intracranial hypertension. The head should be in the midline, because neck rotation or flexion may compromise jugular venous drainage and increase ICP. Head and chest elevation may lower the ICP by enhancing CSF drainage and maximizing cerebral venous output.[7] All patients

Support of the Acutely Failing Liver, 2nd ed., edited by Achilles A. Demetriou and Frederick D. Watanabe.

with intracranial hypertension should be nursed supine with the head and the chest at 30°. However, further elevations to 40° and 60° may paradoxically increase the ICP.[8] Environmental stimulation should be maintained at a minimal level. Most FHF patients with severe encephalopathy are endotracheally intubated to maintain airway protection and adequate ventilation. This may result in ICP fluctuations. Moreover, positive end-expiratory pressure (PEEP) may increase the ICP when mean airway pressures are increased and should be used carefully if at all.[7] Straining against the mechanical ventilator may increase intrathoracic pressure and reduce venous outflow from the head. Coughing, which is a frequent reflex to tracheal tube aspiration, should be avoided for the same reason. Therefore, if necessary, patients are sedated and/or paralyzed with nondepolarizing neuromuscular blockers.

In non-FHF patients, hyperventilation results in ICP reduction through vasoconstriction of the brain blood vessels.[7] The duration of this ICP reduction varies, and the ICP usually returns to the original baseline values within hours of initiation of hyperventilation. The effect of hyperventilation on intracranial hypertension in FHF is not clear.[9] However, most FHF patient management protocols include hyperventilation as one of the first steps in the management of elevated ICP.[1] A prospective, randomized study needs to be performed to establish the effectiveness of hyperventilation in intracranial hypertension secondary to FHF. In the absence of reports of deleterious effects, we continue to use hyperventilation in FHF patients and try to maintain the arterial PCO_2 between 25 and 30 mmHg.

Use of intravenous mannitol improved survival and decreased the ICP level in a controlled trial.[6] Mannitol increases blood osmolarity, thereby inducing fluid movement from brain to blood. Therefore, the efficacy of mannitol depends on an intact blood-brain barrier. Additionally, it cannot be used when renal function is impaired. We use an intravenous bolus of mannitol (0.5-1 g/kg every 6 hours) if needed and if blood osmolarity is less than 310 mOsm/l. In order to maintain high osmolarity, we also may use fluid restriction or administer intravenous furosemide. Acute renal failure is treated with continuous hemofiltration rather than daily hemodialysis. The latter may induce a rise in ICP and a reduction in serum osmolality during the first hour of treatment.[10] In addition, a daily dialysis regimen is more likely to cause episodes of hypotension with further reduction in CPP.

Although there is no controlled study on the effects of intravenous sedation in FHF patients, it is used in the management of their intracranial hypertension. Intravenous thiopental use was studied in a noncontrolled trial in 13 FHF patients and ICP reduction was described in all of them.[11] We prefer use of intravenous bolus administration of fentanyl (0.25-1 mg) and pentobarbital (3-5 mg/kg) to infusion of these agents. Corticosteroids have not proven to be of any benefit in the treatment of cerebral edema and their use has not resulted in improved survival of FHF patients.[11]

Liver transplantation has emerged as the most important advance in the treatment of FHF.[1,12-15] To date, transplantation of a functioning graft is the best treatment for achieving control of brain edema and intracranial hypertension. For this reason, every FHF patient should be referred to a transplant center and considered for OLT as soon as possible. However, some FHF patients in Stage 4 encephalopathy develop severe cerebral injury or brain death during the perioperative period, and these complications are believed to be secondary to perioperative ICP elevation or CPP reduction. In a recent study, 13 among 116 patients (11.2%) who underwent OLT for FHF developed brain death either during or after the procedure, and 2 others suffered significant neurological sequelae.[13] Because of this high rate of neurologic complications, availability of OLT in FHF patients is considered critical. To better understand the ICP changes during OLT in FHF patients, we reviewed the perioperative ICP evolution in 14 consecutive FHF patients recently transplanted at our institution (Table 4.1). Our results demonstrated that the dissection phase of the operation and the period of graft reperfusion were the intervals during which patients were more likely to develop elevations of ICP and that the anhepatic phase was the period least likely to be associated with ICP elevations. Similar findings were reported in the literature a few years ago in a study involving 6 patients.[16] Intracranial hypertension may, however, also occur after OLT: Among 14 patients studied at our Unit, 3 developed episodes of ICP elevation within the first 48 hours postoperatively. This suggests that FHF patients need to be closely monitored in the early postoperative period.

Experimental Treatments

Hypothermia

The effects of hypothermia on ICP in FHF patients have not been adequately studied. Recently, a prospective and randomized study proved the efficacy of moderate hypothermia (34.5°C) in lowering ICP and improving the neurologic outcome of patients suffering from traumatic brain injury.[17] The mechanism(s) of the effects of hypothermia on the brain is not clear. Hypothermia could create a decrease in cerebral metabolism and/or reduce the extracellular concentration of neurotransmitters, particularly glutamate. Hypothermia may limit the acute inflammatory response, preserve the brain-blood barrier and reduce the amount of cytokines released in the brain.[17] In an experimental animal model of FHF (hepatic devascularization in rats), hypothermia lowered brain edema measured by a gravimetric technique and prolonged animal survival.[18] In our laboratory, we investigated the effects of hypothermia on ICP and survival in a large experimental animal model of FHF. Adult pigs underwent total hepatic ischemia by end to side portocaval shunt, ligation of the hepatic artery and transection of all ligaments to the liver. In all animals, a subdural bolt was inserted for continuous ICP monitoring. Postoperatively, animals were kept in a temperature-controlled environment at 21°C. In 8 pigs, no attempts were made to warm the animals (hypothermic group). Another 8 pigs were maintained at 36-38°C throughout the postoperative period (normothermic group). All pigs received glucose (5 g/hr intravenously) as the only supportive measure. In the hypothermic group, core body temperature decreased to 26.5°C at 24 hours. In the normothermic group, core body temperature was greater than 36°C throughout the study. Hypothermic pigs remained hemodynamically stable until 1-2 hours prior to their death. In contrast, normothermic pigs became hypotensive as early as 14 hours postoperatively. The hypothermic group had a significantly longer survival time than the normothermic group. The ICP remained normal throughout the experimental period in 50% of the hypothermic pigs and the remaining animals developed only transient ICP elevation. In contrast, all normothermic pigs developed significant intracranial hypertension. Clinically, we use moderate hypothermia in most of our FHF patients to help treat intracranial hypertension. In two patients this was combined with total hepatectomy; our management algorithm has been described previously.[19] However, hypothermia is not an established treatment of brain edema, and a controlled clinical trial is needed to determine its proper role in the management of the disease. There is, however, general agreement that fever must be treated aggressively in FHF patients, because hyperthermia may induce ICP elevation by increasing cerebral metabolism, blood flow and edema.

Total Hepatectomy

Removal of the necrotic liver has been proposed as a means of achieving control of refractory intracranial hypertension.[20,21] This extreme treatment is based on the "toxic liver syndrome" hypothesis in the etiology of brain edema which suggests that removal of the toxic liver may remove the origin of "toxic products" responsible for edema. Some successful cases have been described in the literature, with clinical improvement of the

Table 4.1. Characteristics of 14 FHF patients recently transplanted at Cedars-Sinai Medical Center

Patient	Gender	Age (yr)	Etiology	Jaundice-Encephalopathy Interval	Head CT	Initial ICP (mmHg)	Highest pre-OLT ICP (mmHg)	Corresponding pre-OLT CPP (mmHg)	Outcome
1	Male	10	Indeterminate	5 days	Edema	17	17	73.5	OLT, Survival
2	Female	18	Acetaminophen	3 days	Edema	7	32	65	OLT, Survival
3	Male	34	Indeterminate	7 weeks	Edema	35	35	38.5	OLT, Survival
4	Male	24	Indeterminate	2 weeks	Normal	15	15	85	OLT, Survival
5	Female	50	Acetaminophen	2 days	Normal	35	52	54	OLT, Survival
6	Male	49	Hepatitis B	2 weeks	Normal	5	77	15	OLT, Survival
7	Female	31	Indeterminate	2 weeks	Edema	45	64	31	OLT, Survival
8	Female	52	Indeterminate	5 weeks	Normal	5	50	60.5	OLT, Survival
9	Female	34	AFLP	3 days	Normal	2	12	89	OLT, Survival
10	Male	51	Indeterminate	8 weeks	Normal	11	15	71	OLT, Late Death

Table 4.1. Characteristics of 14 FHF patients recently transplanted at Cedars-Sinai Medical Center (cont'd)

Patient	Gender	Age (yr)	Etiology	Jaundice-Encephalopathy Interval	Head CT	Initial ICP (mmHg)	Highest pre-OLT ICP (mmHg)	Corresponding pre-OLT CPP (mmHg)	Outcome
11	Female	19	Acetaminophen	4 days	Edema	22	49	20.5	OLT, Survival
12	Female	56	Hepatitis B	1 week	Edema	6	20	49	OLT, Survival
13	Female	27	Indeterminate	1 week	Normal	30	30	59	OLT, Survival
14	Male	26	Indeterminate	3 weeks	Normal	14	40	33	OLT, Early death

AFLP = acute fatty liver of pregnancy
CT = computed tomography
CPP = cerebral perfusion pressure
FHF = fulminant hepatic failure
ICP = intracranial pressure
OLT = orthotopic liver transplant

patients' status after total hepatectomy.[20,21] However, these reports do not include any information about ICP evolution. The use of total hepatectomy in FHF is controversial and needs to be studied in the experimental laboratory. In an anhepatic pig model developed in our laboratory, animals became comatose without a concomitant increase in ICP, whereas animals with ischemic liver failure developed coma and ICP elevation; this suggests that the necrotic liver may be responsible for the elevated ICP in FHF. In our retrospective study of 14 patients with FHF who were transplanted and had subdural ICP monitors, the mean ICP was lower during the anhepatic phase of the OLT than during the preclamp, reperfusion and postoperative phases (Fig. 4.1). As indicated above, in our institution, total hepatectomy was performed in two FHF patients with refractory ICP elevation. In the first case, the anhepatic phase lasted 14 hours and during this time the patient was kept in moderate hypothermia and was treated with a liver support system.[19] The ICP remained normal during the whole anhepatic period. In another patient transplanted for FHF, the anhepatic phase lasted 9 hours. An end to side portocaval shunt was performed, the abdominal wall was closed and the patient stayed in the operating room while waiting for a graft. This patient had intracranial hypertension exceeding 50 mmHg before the surgical procedure. The ICP level normalized during the anhepatic phase and decreased from 21 mmHg-10 mmHg before graft reperfusion. The hemodynamic parameters were stable throughout the anhepatic phase. Both patients recovered fully after successfully undergoing OLT.

Ex Vivo Perfusion of Whole Liver

Ex vivo perfusions of whole animal liver have been attempted to bridge FHF patients to OLT and/or to liver recovery.[22,23] Clinical improvement of the patients' neurologic status has been reported during perfusion; however, despite one isolated case of improvement in ICP, there is no objective evidence that ex vivo perfusion ameliorates brain edema in FHF patients. In fact, the duration of the positive effects of the xenoperfusion is limited by the

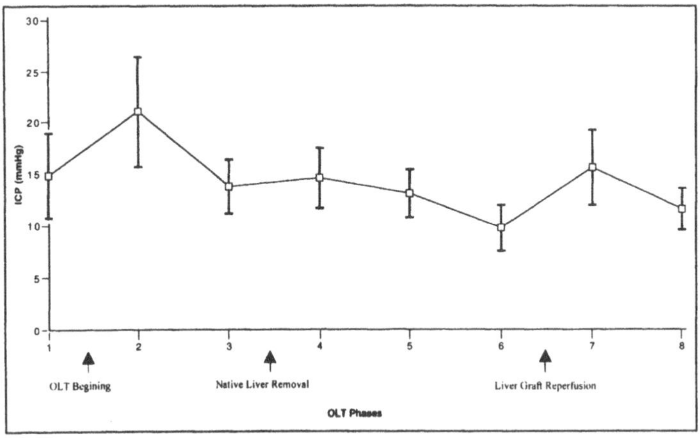

Fig. 4.1. ICP evolution during OLT. CP, Intracranial pressure; OLT, orthotopic liver transplantation. (1) Maximal ICP during the 2 hours before OLT; (2) Maximal ICP during the dissection phase; (3) Last ICP before removal of the necrotic liver; (4) Maximal ICP in the first 15 min of the anhepatic phase; (5) Maximal ICP between 15 min and 1 hour of anhepatic phase; (6) Last ICP before graft reperfusion; (7) Maximal ICP level in the first 15 min following the reperfusion; (8) Maximal ICP level during the completion of the procedure.

dysfunction of the animal liver due to hyperacute rejection secondary to pre-existing xenoantibodies. In the future, use of livers from transgenic pigs resistant to hyperacute rejection may result in improved results.[24]

Hepatocyte Transplantation

Transplantation of isolated hepatocytes has been shown to provide metabolic support and improve survival in various experimental animal models of acute liver failure (90% hepatectomy, D-galactosamine, dimethylonitrosamine, carbon tetrachloride, ischemia).[25-29] Hepatocyte transplantation has also been shown to improve chronic encephalopathy induced by an end to side portocaval shunt in rats.[30] Recently, we have demonstrated in an experimental model of FHF induced by a combination of liver resection and necrosis, that animal survival can be improved by intrasplenic transplantation of a small number of hepatocytes (2% of the native liver mass).[31]

Based on the above observations, we hypothesized that hepatocyte transplantation may prevent, or at least delay, development of intracranial hypertension in a large animal (pig) model of FHF. An experimental group of pigs underwent intrasplenic transplantation of 2.5×10^9 fresh isolated allogeneic hepatocytes under cyclosporine A immunosuppression. After three days, to allow cell engraftment, animals underwent end to side portocaval shunt and transection of all ligamentous liver attachments, including the hepatic artery and bile duct. Control animals received no cells prior to induction of liver necrosis. In all animals, a subdural bolt was inserted for ICP monitoring and femoral vessels were cannulated for blood pressure monitoring and continuous glucose infusion (dextrose 5% in lactated Ringer's solution, 2-3 ml/min). Postoperatively, core body temperature was maintained at/or above 37°C through external heating. Transplanted pigs survived significantly longer than nontransplanted controls. Nontransplanted animals developed severe intracranial hypertension as early as 14 hours postoperatively. In contrast, transplanted pigs maintained normal ICP throughout the postoperative period. At the time of death, the mean ICP in transplanted pigs was 12.0 ± 3.4 mmHg, whereas mean ICP in controls was 29.0 ± 4.0 mmHg ($p < 0.05$). Both groups had similar endpoint core body temperature, heart rate, mean arterial pressure, as well as aspartate aminotransferase, lactic dehydrogenase, albumin, prothrombin time, lactate and arterial blood gases values. Compared to control animals, transplanted pigs had lower blood ammonia levels, lower alkaline phosphatase, lower creatinine and higher total bilirubin. At autopsy, all animals had extensive liver necrosis without signs of other surgical complications (e.g., bleeding, shunt obstruction). In transplanted pigs, splenic sections displayed numerous viable hepatocytes (hematoxylin-eosin, PAS) arranged in clusters and cords, preferentially located in perivascular areas. Although the transplanted pigs eventually died of multiple organ failure, intrasplenic transplantation of allogeneic hepatocytes prevented development of intracranial hypertension in this group. In contrast, nontransplanted pigs showed signs of brain herniation at the time of death. This was the first experimental demonstration of a direct effect of hepatocyte transplantation on ICP in FHF.

In two clinical reports, 12 patients were transplanted with a very small number of hepatocytes (0.01%-0.4% of the liver mass) which were infused either intraperitoneally or intrasplenically.[32,33] Although both studies reported improvement in neurologic status and survival of transplanted patients, the limited number of patients and lack of appropriate controls do not allow reliable conclusions to be reached. More experiments in large animal models are needed in order to investigate the "neuroprotective" potential of transplanted hepatocytes. In addition, three major problems need to be solved before clinical application of hepatocyte transplantation can be established:

1. How to harvest and store a large number of functional hepatocytes from human liver;
2. How to safely carry out transplantation of a significant amount of hepatocytes (at least 5% of the liver mass) considering the anatomic limitations and the severe metabolic disturbances of FHF patients (e.g., coagulopathy); and
3. How to determine the optimal timing of hepatocellular transplantation in the course of FHF.

Extracorporeal Liver Support

Various liver support systems have been developed to treat hepatic encephalopathy and to prevent, arrest, or reverse the development of cerebral edema. These will be discussed in detail in later sections of this monograph.

Summary

Intracranial hypertension has become a major concern in FHF management. Most of the deaths occurring in FHF patients are related to a rise in ICP. The etiology of brain edema is still unknown. An osmotic effect of glutamine accumulation in astrocytes has been proposed; several experimental animal models support this hypothesis. Toxic products from the necrotic liver also have been implicated in the genesis of brain edema in FHF patients. Determining the mechanisms of brain edema in FHF should lead to a specific treatment of intracranial hypertension, which is still lacking. Standard care of intracranial hypertension in FHF patients includes hyperventilation, elevation of the head, intravenous mannitol and sedation. Various experimental systems have been developed to treat cerebral edema in FHF and are currently being tested in clinical trials. Their efficacy remains to be determined. Similarly, the efficacy of total hepatectomy, ex vivo whole liver xenoperfusion and hepatocyte transplantation has not been proven. Improvements in the management of intracranial hypertension should result in better outcomes in FHF patients.

References

1. Hoofnagle JH, Carithers RL, Shapiro C et al. Fulminant hepatic failure: Summary of a workshop. Hepatology 1995; 21:240-252.
2. Lucké B. The pathology of fatal epidemic hepatitis. Am J Pathol 1944; 20:471-525.
3. Ware AJ, D'Agostino AN, Combes B. Cerebral edema: A major complication of massive hepatic necrosis. Gastroenterology 1971; 61:877-884.
4. Gazzard BG, Portmann B, Murray-Lyon IM et al. Causes of death in fulminant hepatic failure and relationship to quantitative histological assessment of parenchymal damage. Q J Med 1975; 176:615-626.
5. Ede RJ, Williams R. Hepatic encephalopathy and cerebral edema. Sem Liver Dis 1986; 6:107-118.
6. Canalese J, Gimson AE, Davis C et al. Controlled trial of dexamethasone and mannitol for the cerebral edema of fulminant hepatic failure. Gut 1982; 23:625-629.
7. Bingaman WE, Frank JI. Malignant cerebral edema and intracranial hypertension. Neurol Clin 1995; 13:479-509.
8. Davenport A, Will EJ, Davison AM. Effect of posture on intracranial pressure and cerebral perfusion pressure in patients with fulminant hepatic and renal failure after acetaminophen self-poisoning. Crit Care Med 1990; 18:286-289.
9. Ede RJ, Gimson AE, Bihari D et al. Controlled hyperventilation in the prevention of cerebral oedema in fulminant hepatic failure. J Hepatol 1986; 2:43-51.
10. Davenport A, Will E, Davison AM. Early changes in intracranial pressure during haemofiltration treatment in patients with grade IV hepatic encephalopathy and acute oliguric renal failure. Nephrol Dial Transplant 1990; 5:192-198.

11. Forbes A, Alexander GJ, O'Grady JG et al. Thiopental infusion in the treatment of intracranial hypertension complicating fulminant hepatic failure. Hepatology 1989; 10:306-310.
12. Devictor D, Desplanques L, Debray D et al. Emergency liver transplantation for fulminant hepatic failure in infants and children. Hepatology 1992; 16:1156-1162.
13. Ascher N, Lake JR, Emond JC et al. Liver transplantation for fulminant hepatic failure. Arch Surg 1993; 128:677-682.
14. Brems JJ, Hiatt JR, Ramming KP et al. Fulminant hepatic failure: The role of hepatic transplantation as primary therapy. Am J Surg 1987; 154:137-141.
15. Bismuth H, Samuel D, Castaing D et al. Orthotopic liver transplantation in fulminant and subfulminant hepatitis. Ann Surg 1995; 222:109-119.
16. Keays R, Potter D, O'Grady J et al. Intracranial and cerebral perfusion pressure changes before, during and immediately after orthotopic liver transplantation for fulminant hepatic failure. Q J Med 1991; 79:425-433.
17. Marion DW, Penrod LE, Kelsey SF et al. Treatment of traumatic brain injury with moderate hypothermia. N Engl J Med 1997; 336:540-546.
18. Traber PG, DalCanto M, Ganger DR et al. Effect of body temperature on brain edema and encephalopathy in the rat after hepatic devascularization. Gastroenterology 1989; 96:885-891.
19. Rozga J, Podesta L, LePage E et al. Control of cerebral edema by total hepatectomy and extracorporeal liver support in fulminant hepatic failure. Lancet 1993; 342:898-899.
20. Ringe B, Lübbe N, Kuse E et al. Total hepatectomy and liver transplantation as two-stage procedure. Ann Surg 1993; 218:3-9.
21. Henderson A, Webb I, Lynch S et al. Total hepatectomy and liver transplantation as a two-stage procedure in fulminant hepatic failure. Med J Aust 1994; 161:318-319.
22. Chari RS, Collins BH, Magee JC et al. Brief report: Treatment of hepatic failure with ex vivo pig-liver perfusion followed by liver transplantation. N Engl J Med 1994; 331:234-237.
23. Parbhoo SP, James IM, Ajdukiewicz A et al. Extracoporeal pig-liver perfusion in treatment of hepatic coma due to fulminant hepatitis. Lancet 1971; ii:659-665.
24. Platt JL. Approaching the clinical application of xenotransplantation. Am J Med Sciences 1997; 313:315-321.
25. Demetriou AA, Reisner A, Sanchez J et al. Transplantation of microcarrier-attached hepatocytes into 90% hepatectomized rats. Hepatology 1988; 8:1006-1009.
26. Hiari S, Kasai S, Mito M. Encapsulated hepatocyte transplantation for the treatment of D-galactosamine induced acute hepatic failure in rats. Eur Surg Res 1993; 25:193-202.
27. Rozga J, Moscioni AD, Demetriou AA. Hepatocyte transplantation: Liver transplantation microchip? In: Sher LS, Makowka L, eds. Intra-abdominal organ transplantation 2000. Austin: R.G. Landes, 1994: 135-142.
28. Takeshita K, Ishibashi H, Kodama M. Hepatocellular transplantation for metabolic support in ischemic liver failure in rats. Transplant Proc 1992; 24:3022.
29. Sandbichler P, Then P, Vogel W et al. Hepatocellular transplantation into the lung for temporary support of acute liver failure in the rat. Gastroenterology 1992; 102:605-609.
30. Ribeiro J, Nordlinger B, Ballet F et al. Intrasplenic hepatocellular transplantation corrects hepatic encephalopathy in portocaval-shunted rats. Hepatology 1992; 15:12-18.
31. Eguchi S, Kamlot A, Hewit WR et al. Hepatocyte transplantation facilitates liver regeneration in rats with fulminant hepatic failure. Surg Forum 1996; 47:445-458.
32. Habibullah CM, Syed HI, Qamar A et al. Human fetal hepatocyte transplantation in patients with fulminant hepatic failure. Transplantation 1994; 58:951-952.
33. Strom SC, Fisher RA, Thompson MT et al. Hepatocyte transplantation as a bridge to orthotopic liver transplantation in terminal liver failure. Transplantation 1997; 63:559-569.

CHAPTER 5

Medical Management of Acute Liver Failure

Frederick D. Watanabe, Elaine Kahaku, Achilles A. Demetriou

Despite advances in critical care management, prognosis in fulminant liver failure (FHF) remains poor, with mortality between 70 and 90%.[1] Introduction of orthotopic liver transplantation has improved outcome and increased patient survival. The paradigm of medical management of FHF focuses on patient stabilization and expectant management until either the patient's liver recovers or liver transplantation can be carried out. In this chapter, guidelines for the medical management of FHF patients will be discussed based primarily upon our own experience at the Liver Support Unit at Cedars-Sinai Medical Center in Los Angeles.

Patient Environment and Monitoring

Successful management of FHF requires near-fanatical diligence, accurate monitoring and rapid response to critical clinical situations. Rapid changes in a patient's status can occur without warning, necessitating early aggressive intervention. This degree of intensive monitoring and capacity to provide early, decisive therapeutic intervention, requires placement of patients in a specialized intensive care unit setting. In general, the clinical syndrome progresses into multisystem involvement requiring a coordinated multidisciplinary team approach to patient management.[2,3] The complexity of the disease necessitates the involvement of adult and pediatric critical care nurses, hepatologists, transplant surgeons, transplant coordinators, intensivists, anesthesiologists, pulmonologists, respiratory therapists, nephrologists, cardiologists, neurologists, neurosurgeons, rehabilitation medicine specialists, infectious disease specialists, pathologists, radiologists and other experts as needed. An established group of physicians, nurses and technicians work as a team to manage all patients with liver failure under strict protocol.

The Liver Support Unit at our institution has been developed to provide a setting where coordinated care using state of the art facilities is carried out. All FHF patients are admitted to the Unit. All diagnosis and management is done by protocol and following clinical pathways developed by members of the Unit. All nursing and technical personnel involved in the care of these patients have received in-service training and become familiar with the diagnostic and treatment protocols. Patient monitoring is carried out using a bedside computerized system. All collected data are stored to allow future review. Therapeutic interventions are similarly recorded and stored as well. Such a system has two major advantages: It serves as a tool allowing accurate monitoring, data retrieval and patient severity of illness stratification and it provides a mechanism to accurately capture all therapeutic interventions. This information has resulted in the development of a useful database for scientific analysis.

Support of the Acutely Failing Liver, 2nd ed., edited by Achilles A. Demetriou and Frederick D. Watanabe.

Upon admission to the Unit, monitoring measures—including placement of a pulmonary artery catheter for cardiovascular pressure and mixed oxygen measurements and arterial line, continuous pulse oximetry, urinary catheter and nasogastric tube placement—are instituted. Intracranial pressure monitoring is initiated in the patients who are comatose and require airway protection. All patients undergo frequent neurologic assessment. Clinical bedside assessment by nurses and physicians is carried out frequently. Subtle changes in neurologic examination, cardiovascular performance, renal function and pulmonary status are monitored and treated early to avoid having the patient experience broad swings in physiologic performance.

The patient is placed in a quiet private room, with a positive pressure environment for reverse isolation if possible. Noise in the room is kept at a minimum and room lighting is dimmed to avoid agitating the patient. The number of personnel and visitors in the room is kept to a minimum. We believe this to be critical in deep hepatic encephalopathy, where increases in ICP can be triggered by minor external stimuli and are difficult to treat.

Vital signs, intravenous input and body fluid output are recorded hourly. As above, we utilize continuous, real-time monitoring which allows instantaneous assessment of patient performance. Significant changes in arterial line blood pressures are verified with manual blood pressure cuff measurement. Nasogastric tube and stool outputs are checked frequently for pH and blood and blood only respectively. The patient's bed has a built-in scale enabling weighing with minimal disturbance to the patient. Neurologic assessment, including mental status and cognition evaluation, is performed regularly. Scheduled repositioning and involvement of a specialized skin care team is important in the obtunded or comatose patient to prevent development of pressure sores and decubitus ulcers. Excessive patient movement, however, is dangerous in the setting of elevated ICP, and in a number of instances we have accepted a high risk of sore and ulcer development to preserve neurologic function and treated the ulcers following transplant or recovery.

Fluids, Electrolytes and Nutrition

Fluid management in FHF patients requires maintenance of a balance to avoid fluid overload as well as dehydration, as multiorgan failure develops. Central pressure monitoring, cardiac performance and peripheral resistance are determined using a pulmonary artery catheter. Renal failure can occur in the face of pulmonary edema and peripheral vascular collapse; strict monitoring of fluid balance is therefore required. Furthermore, cerebral edema and intracranial hypertension require careful fluid administration to avoid expansion of the intravascular space and exacerbation of cerebral edema. One should take into account "hidden" fluid administration, including fluid used as a vehicle for administering intravenous medications, fluid given to maintain catheter patency, insensible losses and third spacing. Measured fluid balance should be verified twice daily by body weight determinations to ensure accuracy of the replacement estimates. Fluid replacement strategies should anticipate changes in losses (developing renal failure, fever, etc.) and the need for administration of blood products. Rapid bolus fluid administration should be avoided in favor of anticipatory adjustments in fluid delivery so as not to aggravate pulmonary edema, heart failure or intracranial hypertension.

Electrolyte and acid-base imbalances frequently occur in FHF. The magnitude of disturbance and difficulty in management depends upon the extent and severity of multiorgan involvement. Most commonly seen are hyponatremia, hypokalemia, hypocalcemia, hypophosphatemia and hypomagnesemia. Calcium, phosphorus and magnesium should be supplemented as needed. These deficiencies may be exacerbated by gastrointestinal and renal losses. Repeated transfusions may inadvertently exacerbate hypocalcemia, since sodium citrate is used as an anticoagulant in fresh frozen plasma, cryoprecipitate and packed red cells. The failing liver does not metabolize citrate, allowing the formation of calcium citrate complexes. Children and small adults, by having relatively smaller total blood volumes, are at higher risk for this complication.

Sodium and potassium supplementation must be accompanied by frequent accounting of all electrolyte intake sources and losses. Overzealous sodium replacement may significantly increase total body sodium and stimulate ascites formation. Hyperkalemia can result in cardiac arrhythmias. Frequent serum and urine (if possible) measurements are needed to assess true losses. Sodium and potassium in medications or blood products can contribute significant amounts to body stores. Hepatic necrosis can cause unexpected potassium release as well as profound metabolic acidosis.

Acidosis in FHF results from increased lactic acid production and decreased hepatic handling of lactate by the failing liver. Compensatory respiratory alkalosis is present early, but if encephalopathy progresses, respiratory acidosis may result. Renal losses of bicarbonate can cause metabolic acidosis. Gastrointestinal losses of acid by nasogastric tube or bicarbonate loss through diarrhea may further complicate the situation. Sodium or potassium bicarbonate infusions should be used in severe acidosis—with consideration given to any anion species derangement. Acetate provides twice the bicarbonate load and is metabolized outside the liver. Thus continuous infusion of acetate salts can be utilized if sodium and potassium restrictions are severe.

The role of nutritional intervention in FHF is not clear. Individual circumstances dictate the aggressiveness of nutritional supplementation. In the FHF patient, the impact of nutritional support other than maintenance of euglycemia is unknown. Hypoglycemia is present in up to 40% of patients with FHF.[5] The etiology is multifactorial and includes serum insulin elevation from decreased hepatic clearance, impaired hepatic gluconeogenesis and impaired release of glucose into the circulation. In general, intravenous infusion of dextrose at 4 mg/kg/min offers a reasonable glucose load without the risk of hyperglycemia.[6] A 10% dextrose solution is usually an adequate vehicle; however, higher concentrations of glucose should be considered with continued hypoglycemia and/or fluid restriction.

Hyperglycemia can contribute to immunosuppression by impairing phagocytosis and may result in osmotic diuresis, complicating fluid management. Serum glucose monitoring should be carried out hourly in the comatose patient and every 4 hours in those who are awake. Hypoglycemia should be corrected rapidly by 50% dextrose infusion followed by increasing the infusion rate of dextrose or increasing the concentration of dextrose as needed. Caloric supplementation has not been extensively studied in FHF, though it appears that glucose infusion alone may not be optimal. Caloric requirements should be determined by indirect calorimetry if available; however, patients rarely require more than 3000 kcal/day. Amounts in excess of this may not be beneficial.[6]

Amino acids should be withheld in the initial management of FHF to avoid excessive nitrogen load. Limited nitrogen supplementation (70-80 g of protein/day) may be provided by the parenteral route, though no earlier than the third day of disease.[4] In general, in these patients we do not consider protein supplementation until the seventh day of nutritional deprivation. This protein load results in negative nitrogen balance. Individuals with multiorgan failure require 1.6-2.5 gm/kg/d of protein.[6] Additionally, FHF patients clinically demonstrate a high degree of catabolism leading to hypoalbuminemia, lymphopenia and anergy.[7] Paradoxically, skeletal muscle catabolism releases amino acids and contributes to an increased plasma ammonia load. Increasing nitrogen loads may offset the catabolic effect at the risk of exacerbating hepatic encephalopathy.

The type of proteins and amino acids delivered may be significant in the FHF patient. Hepatic encephalopathy may be influenced by concentrations of ammonia and aromatic amino acids. a-Keto analogs of amino acids are transaminated in the intestinal mucosa into amino acids without supplying additional nitrogen. Limited studies demonstrate that keto-acids are efficiently converted into amino acids and incorporated into protein, maintaining nitrogen balance and reducing ammonia levels in cirrhotic patients.[8,9] Parenteral solutions of branched-chain amino acids have also been utilized with inconclusive results.[10] Though promising, specialized parenteral formulations should be restricted to use under investigational protocols.

Hepatic Encephalopathy and Cerebral Edema

The presence of hepatic encephalopathy and cerebral edema significantly influences the management and outcome of FHF. Neurologic deterioration with irreversible brain damage is a contraindication for liver transplantation in up to 30% of patients.[1] Neurologic complications represent a major cause (30%) of posttransplantation mortality.[1] Autopsy studies demonstrate cerebral edema in 80% of patients who die from FHF.[11,12] The pathogenesis of encephalopathy and increased intracranial hypertension have been discussed in previous sections of this monograph.

Diagnosis of hepatic encephalopathy requires the recognition and correction of other disorders affecting cerebral function. Electrolyte and fluid abnormalities, hypoglycemia, azotemia, metabolic acidosis or alkalosis, hypoxia and plasma hyperosmolality need to be corrected. Sedatives and paralytic agents should be avoided during the initial assessment if possible. If sedation, or paralysis, is required, the combination of poor hepatic function and shunting of blood away from the liver may greatly lengthen the drug elimination curve, complicating the interpretation of patient assessment.

The classic therapeutic objectives in hepatic encephalopathy have been to minimize ammonia formation, augment ammonia elimination and correct factors, or avoid conditions, which may result in deterioration of mental status and elevation of intracranial pressure. Several studies have demonstrated that increasing serum ammonia levels can induce hepatic encephalopathy.[13,14] Conversely, decreasing serum ammonia can ameliorate encephalopathy.[15,16] Ammonia levels, however, do not correlate with the severity of encephalopathy nor are they predictive of development of encephalopathy.[4] Additionally, ammonia is neuroexcitatory and central nervous system depression is more commonly seen in hepatic encephalopathy.[17]

The colon is the primary target for enhancing ammonia elimination and reducing ammonia synthesis. Antibiotics and lactulose can be administered early in the course of the disease, though their efficacy in FHF-associated hepatic encephalopathy is not as established as in chronic liver disease. Lactitol, sorbitol and lactose (in the lactose-intolerant patient) have been used in place of lactulose.[4] Neomycin is a minimally absorbed antibiotic commonly used for gut decontamination. Neomycin is associated with ototoxicity and nephrotoxicity and thus should not be administered for long periods of time. Other oral antibiotics include polymyxin B, metronidazole and vancomycin. "Gut Decontamination Solutions" are home-grown cocktails of these antibiotics and serve the same purpose. Parenterally delivered antibiotics excreted in the bile can also be used to reduce the gastrointestinal bacterial load.

Lactulose is a synthetic disaccharide cathartic that can be delivered by oral, nasogastric or high enema route. Dosage should be titrated to achieve 2-4 loose bowel movements per day. Lactulose is neither absorbed nor metabolized in the upper gastrointestinal tract. Upon reaching the colon, bacterial degradation acidifies the luminal contents and causes an intraluminal osmotic shift. This acidification inhibits coliform bacterial growth, thereby reducing ammonia production. Additionally, low intraluminal gut pH results in the conversion of ammonia to ammonium ions that do not easily enter the bloodstream. The cathartic action of lactulose clears ammonium ions from the bowel. Since this sugar needs to be digested, failure of lactulose to acidify stool adequately can be a result of antibiotic-induced suppression of colonic bacteria. The simultaneous use of antibiotics and lactulose may be synergistic if stool pH is less than 6; however, with a higher pH, there is decreased coliform bacterial activity and decreased likelihood that the lactulose will be effective. Aggressive lactulose therapy may induce volume depletion and electrolyte imbalances. Metabolic acidosis is a rare occurrence. Additionally, stool acidification can be problematic if a patient is intolerant to stimulation due to high ICP and cannot be moved. Lactulose should be discontinued if patient stimulation due to multiple bowel movements or acidic feces irritation results in an increase in ICP.

Aromatic amino acids are known neurotransmitter precursors; it has been suggested that their products interfere with the activity of true neurotransmitters. Aromatic amino acids are absorbed from dietary protein, and are released by the skeletal muscle during catabolism and from the liver as hepatocytes become necrotic. The plasma ratio of branched-chain to aromatic amino acids decreases steeply during encephalopathy.[18] Aromatic and branched-chain amino acids compete to cross the blood-brain barrier. Increasing the plasma concentrations of branched-chain amino acids favors their transfer into the brain. In general, however, administration of parenteral branched-chain amino acid solutions has not been found beneficial when compared to use of conventional parenteral amino acid solutions.[4]

Neuroinhibitory processes may also contribute to the development of hepatic encephalopathy. The levels of the inhibitory neurotransmitter g-aminobutyric acid (GABA), and its postsynaptic GABA receptor, increased in a rabbit model of liver failure;[19] however, results were inconclusive in patients.[20-22] Autopsy studies have demonstrated the presence of diazepam and desmethyldiazepam in brain tissue, indicating the existence of natural benzodiazepines.[23] Limited trials using flumazenil, a pure benzodiazepine antagonist, suggest that hepatic encephalopathy can be transiently reversed or ameliorated.[24,25] This may allow one to test the neurologic integrity of the comatose patient. However, lack of consistent response and risk of seizure induction as a consequence of flumazenil preclude the routine use of this medication. The pathogenesis of encephalopathy may not be purely a benzodiazepine response, since the postsynaptic GABA receptor is closely linked to the benzodiazepine and picrotoxin (barbiturate) receptors. Thus flumazenil may inhibit both benzodiazepine and GABA effects.

Cerebral edema should be managed aggressively. Neurologic physical examination is not sensitive enough for detecting increased ICP, so clinical suspicion should be maintained.[5] The earliest signs and symptoms suggestive of development of cerebral edema include systolic hypertension and increased muscle tone. Headache, vomiting and papilledema are present inconsistently. Later in its course, cerebral edema presents with decerebrate posturing, hyperventilation, myoclonus, seizures, trismus and opisthotonos.

Computed axial tomography (CAT) scan of the brain is not a sensitive test for detecting early cerebral edema; 25-30% of patients with high ICP have no radiographic changes.[26,27] The CAT scan, however, is useful in ruling out intracranial bleeding. We advocate using ICP monitoring in all comatose patients and patients exhibiting neurologic deterioration. Several types of ICP monitors are available. Intraventricular monitors are the most accurate and precise; however, they also have a high complication rate. Epidural monitors are the safest, but they are often inaccurate. Subdural placement of an ICP monitor has intermediate risk with relatively accurate readings. In one series, subdural ICP monitoring had a 22% incidence of hemorrhage.[27] Recently introduced fiberoptic subdural monitors promise equal, if not better, accuracy with reduced risk of bleeding. All ICP monitors increase the risk of infection, especially meningitis.

Placement of an ICP monitor requires an experienced neurosurgeon and aggressive treatment of concomitant coagulopathy. Fresh frozen plasma infusions are used to bring the prothrombin time below 20 seconds, and platelet transfusions are indicated if the patient is thrombocytopenic (<50,000/mm^3). Additionally, FFP is administered during ICP monitor insertion. Once ICP monitoring is established, FFP bolus administration is repeated as needed to keep the prothrombin time at or below 20 seconds.

Management of increased ICP involves use of hyperventilation, minimal stimulation, head elevation, maintenance of hemodynamic stability and mannitol infusion. Dexamethasone administration has not been shown to have a beneficial effect in FHF and should not be used.[28] Rapid fluid boluses and aggressive, or inappropriate, correction of electrolyte imbalances can aggravate the situation and should be avoided. Fluid restriction should be attempted, though it is frequently circumvented by the volume of infused medications, need to maintain hemodynamic stability and need to administer blood products. The therapeutic goal is to maintain the cerebral perfusion pressure (difference between mean arterial pressure and intracranial pressure) above 50 mmHg. This usually requires maintaining the ICP below 20 mmHg. Sustained low cerebral perfusion pressure (less than 40 mmHg) is associated with poor outcome.[28-30]

Mechanical hyperventilation lowers ICP but has not been shown to decrease morbidity.[3] Cerebral blood flow is linearly related to the partial pressure of carbon dioxide between 20 and 60 mmHg. Lowering carbon dioxide pressure to 25-30 mmHg maximizes cerebral vascular constriction and reduces blood flow. This vascular effect progressively diminishes after six hours of therapy, though clinical response may be apparent for days.

Mannitol infusions have up to an 80% response rate and have been shown to increase survival in FHF patients without renal failure.[23,28] Serum osmolality should be measured frequently and maintained at 300-320 mOsm. Mannitol should be withheld if osmolality is 320 mOsm or higher, if renal failure occurs or if oliguria and rising serum osmolality develop simultaneously. Repeated administration of mannitol may reverse the osmotic gradient. Thus, mannitol should be discontinued if the ICP does not respond after the first few boluses.

Renal Impairment

Renal failure occurs in up to 50% of patients with FHF, often due to hepatorenal syndrome.[1,2,32] Hepatorenal syndrome is the unexplained development of reversible renal failure in both acute and chronic liver disease, characterized by development of azotemia, oliguria, low urinary sodium excretion (<10 mEq/d) and increased urine to serum osmolality ratio in the absence of active urinary sediment. In FHF, azotemia may not be apparent until later, as urea production may be reduced. Hyperkalemia may quickly develop as hepatic injury continues, renal function declines and potassium retention occurs. Hyponatremia, however, may worsen as the amount of free water retention is in excess of sodium retention. Acute tubular necrosis is another cause of renal failure in patients with FHF, especially if a patient has not been resuscitated adequately, or has experienced prolonged hypotension or ingestion of hepatotoxins which are also nephrotoxic (e.g., acetaminophen).

As mentioned earlier, administration of fluid boluses to maintain renal perfusion may induce pulmonary edema or increase the ICP and thus should be used judiciously. Close attention must be given to low cardiac output and blood pooling from peripheral vasodilatation as possible etiologic factors for poor renal perfusion. Adequate urine volumes can be maintained with judicious use of loop diuretics and low dose, i.e., renal dose, dopamine infusion. Depleted intravascular volume should be managed with administration of blood components and/or volume expanders. Since plasma albumin is invariably low, salt-poor albumin solutions may be preferred over a carbohydrate-based volume expander. If oliguria is present, and especially if mannitol is administered to treat ICP, hemodialysis/

hemofiltration may be needed to maintain optimal fluid volume and normal serum osmolality.

Fluid restriction must be instituted once renal failure is established. Sodium and potassium losses should be estimated based on the amount and type of body fluids lost and replaced appropriately. Hyponatremia should not be corrected with hypertonic saline unless the patient has a hyponatremic seizure. Hemofiltration to remove excess free water will resolve the problem without upsetting total body sodium content.

Pulmonary Disease

Pulmonary complications occur frequently in FHF patients. There is particular predisposition to low pressure pulmonary edema and development of adult respiratory distress syndrome.[23] Pulmonary edema is seen in up to 40% of FHF patients.[30] This is believed to be due to pulmonary capillary changes which result in increased permeability, though development of multiorgan failure syndrome cannot be excluded.[2] Patient management is supportive, with monitoring by continuous pulse oximetry, frequent arterial blood gas measurements, administration of supplemental oxygen and use of mechanical ventilation as needed. Sedative and paralytic agents may be required to ensure tolerance of ventilation. These should be used sparingly, as they hinder neurologic evaluation and may have prolonged half lives from decreased hepatic metabolism and/or urinary excretion. Additionally, neurologically depressed patients are at increased risk for aspiration pneumonia, which can result in inability to carry out liver transplantation.

Infection

Infection poses a serious threat to FHF patients, both by placing them at risk for sepsis and by being a contraindication to liver transplantation. Immunologic defects include impaired opsonization, impaired chemotaxis, neutrophil and Kupffer cell function impairment and complement deficiency.[34-37] Bacterial infection is reported to be prevalent in more than 80% of cases, usually with a respiratory or urinary source.[38] Bacteria are seen in 25% of the patients, with *Staphylococcus sp.*, *Streptococcus sp.* and Gram negative rods as the most common pathogens.[5,35,38] Iatrogenic sources must be considered, as most patients have percutaneous lines and an indwelling urinary bladder catheter. In one series, fungal infections were found in a significant number of patients, with *Candida albicans* cultured in 33% of the patients studied.[39] These patients were predominantly in renal failure and had been treated with antibiotics for periods longer than 5 days.

Despite the high prevalence of infection, prophylactic antibiotic administration is not advocated without suspicion of active infection. However, the threshold for starting antibiotics should be low, as the usual clinical presentation with fever and leukocytosis may be absent in 30% of FHF patients.[38] Surveillance cultures for bacteria and fungi must be obtained at frequent intervals from blood (peripheral and central lines), urine, sputum and open wounds. Ascitic fluid should be cultured if ascites is present. Additionally, chest radiographs should be obtained to identify a developing infiltrate. High dose, broad spectrum antibiotics should be started at the first sign of infection, with narrowing of coverage as soon as an organism is identified. Consideration needs to be given to initiation of either amphotericin B or other anti-fungal therapy if there is either a positive fungal culture or fever persisting beyond 5 days, especially in renal failure patients. The duration of antimicrobial therapy should be adjusted to the individual patient's situation. Follow-up cultures are recommended if a specific organism is isolated.

Coagulation and Hemorrhage

Bleeding is a frequent complication of FHF. Coagulopathy results from impaired hepatic synthesis of clotting factors, impaired platelet synthesis and platelet dysfunction or a combination of the above. Gastrointestinal bleeding is the most common clinical manifestation of coagulopathy and can be massive and life-threatening. Intracranial bleeding with neurologic sequelae is another devastating complication of coagulopathy in FHF.

Plasma activity of all the clotting factors synthesized by the liver (II, V, VII, IX and X) is depressed in FHF. Factors II, VII, IX and X are vitamin K-dependent, and factor V is synthesized independently of vitamin K availability. Factor II, with a half life of 2 hours, is the first to be depleted in hepatocellular dysfunction and is also the first to be repleted with hepatocellular recovery. Increased consumption of clotting factors also may occur as a result of development of disseminated intravascular coagulation (DIC).

Thrombocytopenia and abnormalities of platelet function are frequently encountered in FHF. Acute splenomegaly, consumptive coagulopathy and bone marrow suppression all contribute to the development of thrombocytopenia. Conversely, clearance of older platelets from the blood by the reticuloendothelial system is hindered creating an older, less effective platelet pool. Platelet function may be further impaired by abnormal plasma and intraplatelet lipid levels; this results in decreased adhesion and aggregation.[40]

Frequent monitoring of the prothrombin time, partial thromboplastin time, platelet count, fibrinogen level and fibrin split products should be initiated before there is clinical evidence of bleeding. Prophylactic infusion of fresh frozen plasma (FFP) has not been shown to alter outcome in FHF.[5] Thus, the use of FFP should be restricted to the coagulopathic patient. Similarly, prophylactic heparinization for DIC has not been proven to be effective in FHF.

Vitamin K should be administered parenterally following diagnosis of FHF.[41] With wide-spread hepatic damage, however, the prothrombin time may not change significantly. Platelet transfusions have not been shown to prevent bleeding. In one study, however, a mean platelet count of 50,000/mm^3 was associated with a higher incidence of occurrence of gastrointestinal hemorrhage.[40] Presently, we transfuse platelets either for thrombocytopenia (platelet count less than 50,000/mm^3) or in patients who are thrombocytopenic and experiences active bleeding.

The most common consequence of coagulopathy is gastrointestinal bleeding. Though most episodes can be controlled, gastrointestinal bleeding is associated with high morbidity and mortality.[42] Lower gastrointestinal bleeding is associated with a significant increase in plasma ammonia level and may exacerbate hepatic encephalopathy.[4] A nasogastric tube should be placed early in the disease course. Gastric pH needs to be checked frequently and maintained greater than 4.5.[4] H_2 receptor antagonists effectively control high acidity and reduce the incidence of gastrointestinal hemorrhage.[43] Beginning with the lowest therapeutic dose, measuring serial gastric pH and adjusting the dose upwards as needed minimizes the side effects of these medications. Cimetidine has been implicated as having a higher incidence of neurologic side effects, making ranitidine the preferred histamine antagonist in this setting.[42] However, trials comparing various H_2 blockers in FHF patients are lacking. Antacids can augment histamine blockade; however, diarrhea and electrolyte and mineral imbalances can occur. Proton pump antagonists such as omeprazole have not been carefully studied in this setting.

In conclusion, medical management of FHF requires a multispecialty, multidisciplinary approach because of the complexity of the underlying disease. Hemodynamic and respiratory support and prevention and treatment of cerebral edema are the major goals of medical therapy. Patients with rapidly deteriorating liver function require urgent orthotopic liver transplantation, as it remains the only effective therapeutic modality currently available.

References

1. Lidofsky S. Liver transplantation for fulminant hepatic failure. Gastroenterol Clin North Am 1993; 22:257-269.
2. Douglas DD, Rakela J. Fulminant hepatitis. In: Kaplowitz N, Ed. Liver and Biliary Diseases. 1st ed. Baltimore: Williams and Wilkins, 1992: 279-288.
3. Bihari DJ, Gimson AES, Williams R. Cardiovascular, pulmonary and renal complications of fulminant hepatic failure. Semin Liv Dis 1986; 6:119-128.
4. Marsano L, McClain C. How to manage both acute and chronic hepatic encephalopathy. J Crit Illness 1993; 8:579-600.
5. Fingerote RJ, Bain VG. Fulminant hepatic failure. Am J Gastroenterol 1993; 88:1000-1010.
6. Bagley JS, Wan JMF, Georgieff M et al. Cellular nutrition in the support of early multiple organ failure. Chest 1991; 100(3) supp:182s-188s.
7. O'Keefe SJ, El-Zayadi AR, Carraher TE et al. Malnutrition and immunoincompetence in patients with liver disease. Lancet 1980; X:615-617.
8. Maddrey WC, Weber FL, Coulter AW et al. Effects of ketoanalogues of essential amino acids in portal-systemic encephalopathy. Gastroenterology 1976; 71:190-195.
9. Herlog HF, Maddrey WC, Walser M. The use of ornithine salts of branched-chain ketoacids in portal-systemic encephalopathy. Ann Intern Med 1980; 93:545-550.
10. Ferenci P. Critical evaluation of the role of branched-chain amino acids in liver disease. In: Thomas JC, Jones EA, eds. Recent Advances in Hepatology. Edinburgh: Churchill Livingstone Inc. 1986; 137-154.
11. Ware AJ, D'Agostina A, Combs B. Cerebral edema: A major complication of massive hepatic necrosis. Gastroenterology 1971; 61:877-884.
12. Sik DBA, Hanid MA, Trewby PN et al. Treatment of fulminant hepatic failure by polyacrylonitrile membrane hemodialysis. Lancet 1977; ii:1-3.
13. Conn HO, Lieberthal MM. The hepatic coma syndromes and lactulose. Williams and Wilkins, Baltimore, Md. 1979.
14. Gabuzda GL Jr, Philips GB, Davidson CS. Reversible toxic manifestation in patients with cirrhosis oh the liver given cation-exchange resins. N Engl J Med 1952; 248:124-130.
15. Elkington SG, Floch MH, Conn HO. Lactulose in the treatment of chronic portal-systemic encephalopathy. A double-blind clinical trial. N Engl J Med 1969; 281:408-412.
16. Conn HO, Levy CM, Vlahcevic ZR et al. Comparison of lactulose and neomycin in the treatment of chronic portal-systemic encephalopathy: A double-blind controlled trial. Gastroenterology 1977; 72:573-583.
17. Schafer DF, Jones EA. Hepatic encephalopathy. In: Zakim D, Boyer TD, eds. Hepatology: A Textbook of Liver Disease. 2nd Ed. Philadelphia: JB Lippincott Co, 1987; 925-948.
18. Ansley JD, Isaacs JW, Rikkers LF et al. Quantitative tests of nitrogen metabolism in cirrhosis: Relation to other manifestations of liver disease. Gastroenterology 1978; 75:570-579.
19. Jones EA, Schafer DF. Hepatic encephalopathy: A neurochemical disorder. In: Popper H, Schaffer F, eds. Progress in Liver Disease. New York: Frune and Stratton Inc, 1983: 525-539.
20. Minuk GY, Winder A, Burgess EJ et al. Serum gamma-aminobutyric acid (GABA) levels in patients with hepatic encephalopathy. Hepatogastroenterology 1985; 32:171-174.
21. Maddison JE, Dodd PR, Morrison M et al. Plasma GABA, GABA-like activity and the brain: GABA-benzodiazepine receptor complex in rats with chronic hepatic encephalopathy. Hepatology 1987; 7:621-628.
22. Levy LJ, Leek J, Losowsky MS. Evidence for gamma-aminobutyric acid binding in the plasma of humans with liver disease and hepatic encephalopathy. Clin Act 1987; 73:531-534.
23. Basile AS, Hughes KD, Harrison PM et al. Elevated brain concentration of 1,4-benzodiazepines in fulminant hepatic failure. N Engl J Med 1991; 325:473-478.
24. Grimm G, Ferenci P, Katzenschager R et al. Improvement of hepatic encephalopathy treated with flumazenil. Lancet 1988; ii:1392-1394.
25. Bansky G, Meier PJ, Riederer E et al. Effects of the benzodiazepine receptor antagonist flumazenil in hepatic encephalopathy in humans. Gastroenterology 1989; 97:744-750.

26. Munoz SJ, Robinson M, Northrup B et al. Elevated intracranial pressure and computed tomography of the brain in fulminant hepatic failure. Hepatology 1991; 13:209-212.
27. Lidofsky SD. Bass NM, Preger MC et al. Intracranial pressure monitoring and liver transplantation for fulminant hepatic failure. Hepatology 1992; 16:1-7.
28. Canalese J, Gimson A, Davis C et al. Controlled trial of dexamethasone and mannitol for the cerebral edema of fulminant hepatic failure. Gut 1982; 23:625-629.
29. Schafer DF, Shaw BW. Fulminant hepatic failure and orthotopic liver transplantation. Semin Liver Dis 1989; 9:189-194.
30. Inagaki M, Shaw B, Schafer D et al. Advantages of intracranial pressure monitoring in patients with fulminant hepatic failure. (Abstract) Gastroenterology 1992; 102:A826.
31. Ede RJ, Gimson AE, Bihari D et al. Controlled hyperventilation in the prevention of cerebral oedema in fulminant hepatic failure. J Hepatol 1986; 2:41-43.
32. Ring-Larsen H, Palazzo U. Renal failure in fulminant hepatic failure and in terminal cirrhosis: A comparison between incidence, types and prognosis. Gut 1981; 22:585-591.
33. Trewby PN, Warren R, Contini S et al. Incidence and pathophysiology of pulmonary edema in fulminant hepatic failure. Gastroenterology 1978; 74:859-865.
34. Larcher VF, Wyke RJ, Mowat AP et al. Bacterial fungal infection in children with fulminant hepatic failure: Possible role of opsonization and complement deficiency. Gut 1982; 23:1037-1043.
35. Wyke RJ, Canalese JC, Gimson AE et al. Bacteremia in patients with fulminant hepatic failure. Liver 1982; 2:45-52.
36. Bailey RJ, Woolf IL, Cullens H et al. Metabolic inhibition of polymorphonuclear leukocytes in fulminate hepatic failure. Lancet 1976; 1:1162-1163.
37. Imawari M, Hughes RD, Love CD et al. Fibronectic and Kupffer cell function in fulminant hepatic failure. Dig Dis Sci 1985; 30:1028-1033.
38. Rolando N, Harvey F, Brahm J et al. Prospective study of bacterial infection in acute liver failure: An analysis of fifty patients. Hepatology 1990: 11:49-53.
39. Rolando N, Harvey F, Brahm J et al. Fungal infection: A common, unrecognized complication of acute liver failure. J Hepatology 1991; 12:1-9.
40. O'Grady JG, Langley PG, Isola LM et al. Coagulopathy in fulminant hepatic failure. Semin Liv Dis 1986; 6:159-163.
41. Russell GJ, Fitzgerald JF, Clark JH. Fulminant hepatic failure. J Peds. 1987; 111:313-319.
42. Treem W. Hepatic failure. In Walker WA, Durie PR, Hamilton JR et al. eds. Pediatric Gastrointestinal Disease, 1st ed. Philadelphia: BC Decker, 1991: 146-192.
43. MacDougall BR, Bailey RJ, Williams R. H2-receptor antagonists and antacids in the prevention of acute gastrointestinal hemorrhage in fulminant hepatic failure. Lancet 1977; 2:617-619.

CHAPTER 6

Surgical Management of Acute Liver Failure

Walid S. Arnaout, David Reybould

Fulminant hepatic failure (FHF) continues to be one of the most devastating conditions affecting thousands of individuals each year. By definition, these are patients who were previously healthy without any known underlying liver disease. Historically, the outcome of these patients in most cases has been poor. With the introduction of orthotopic liver transplantation in the late seventies and following the NIH consensus conference on liver transplantation in 1983, orthotopic liver transplantation became an accepted therapeutic modality for patients with end stage liver disease and was no longer considered an experimental procedure. Since then, liver transplantation has been used as a treatment modality for patients with FHF, with a significant increase in patient survival and reduction in the associated morbidity.

With the introduction of orthotopic liver transplantation (OLT) as a treatment of choice for irreversible severe acute liver failure, overall patient survival has improved from less than 20% to greater than 50%.[1-9] A recent report described actuarial one year survival of 92% for FHF and 100% for subfulminant hepatic failure (SFHF).[10] As the experience with OLT for FHF increased, it became apparent that appropriate, and optimal, patient selection is essential for a successful outcome. Patients with FHF should be considered for OLT prior to development of irreversible brain injury or multisystem organ failure and sepsis. Patient selection should be based on a clear understanding of the natural history of the disease as well as the underlying etiology and chances of spontaneous recovery without liver transplantation. One of the most difficult aspects in the management of these patients is lack of reliable prognostic indicators or criteria predicting outcome. This means that a small number of patients who could have recovered spontaneously, or have developed irreversible brain damage, undergo unnecessary liver transplantation. This is a significant problem if one considers the severe organ donor shortage as well as the financial and medical consequences of liver transplantation and commitment to life-long immunosuppression.

According to UNOS data, 10-11% of pediatric liver transplants and 6% of adult liver transplants in the US are performed for acute liver failure annually.[11] In both pediatric and adult patient populations non-A, non-B, non-C hepatitis was the most common underlying etiology for liver failure, at 13% and 23% respectively.

Definitions and Patient Selection

As discussed in earlier sections of the monograph, the definition of acute liver failure and its classification into fulminant versus subfulminant depends on the temporal relation

Support of the Acutely Failing Liver, 2nd ed., edited by Achilles A. Demetriou and Frederick D. Watanabe.

between the initial onset of illness and manifestation of jaundice, encephalopathy and coagulopathy. One of the main problems is the difficulty to precisely determine the time of onset of the disease process and thus define the interval between disease onset and establishment of liver failure. The classic definition of Trey and Davidson is based on encephalopathy developing within 8 weeks from the onset of the illness.[12] In a more recent classification, Bernuau[1] defined FHF as acute hepatic failure (AHF) complicated by encephalopathy occurring less than 2 weeks after the onset of jaundice.[3] Recognizing that the prognosis and clinical findings were different in those patients who developed encephalopathy later than 2 weeks, the term subfulminant hepatic failure (SFHF) was introduced to describe AHF complicated by encephalopathy developing between 2 to 12 weeks after the onset of jaundice. Despite the various definitions and classifications of AHF, the fact remains that common to all these definitions is the lack of pre-existing liver disease.

Several prognostic criteria and indicators have been proposed in an attempt to predict the outcome of these patients with optimal medical management. The two major factors determining whether a patient with FHF is likely to survive are the extent of liver necrosis and the degree of hepatocyte regeneration. Clinical, biochemical and radiographic studies assessing these factors, individually or in various combinations, are being used as a logical basis for predicting outcome. An ideal prognostic indicator should be 100% sensitive and specific and the assessment should be completed within few hours of admission to a liver transplant unit. Those patients who are unlikely to recover with aggressive medical therapy should be considered as candidates for emergency liver transplantation.

Over the past twenty years, several prognostic criteria have been proposed to assess the outcome of acute hepatic failure based on retrospective analyses of data from a relatively small number of patients. Christensen analyzed 21 variables in 33 patients with FHF.[14] Few variables were found to be predictive of outcome, including age, male sex, duration of history, hepatitis B, low serum glucose and albumin and prolonged prothrombin time. Discriminant score was calculated based on the predictive coefficient of each variable and survival rate was significantly better ($p = 0.000003$) if the score was greater than 0.5 (86%) compared to that of patients with a score less than 0.5 (5%). Tygstrup and Ranek[1] reviewed the literature on assessment of prognosis in FHF and examined the predictive value of several variables with a multivariant analysis. Despite the large number of prognostic variables studied in FHF, this retrospective analysis did not identify useful outcome predictors among patients with FHF. Despite lack of statistical significance in the predictive value of single variables, several studies have shown a number of variables to be strongly related to prognosis. These include:

1. Etiology of liver disease, such as acute hepatitis A or B infection or acetaminophen toxicity;
2. Duration between onset of disease and encephalopathy;
3. Degree of encephalopathy;
4. Age of the patient; and
5. Presence of cerebral edema.

Following the review by Tygstrup and Ranek,[1] O'Grady at the King's College Hospital in London completed a univariant and multivariant analysis of data from 588 patients with acute liver failure managed at their unit between 1973-1985.[15] In a multivariant analysis, underlying etiology was found to be the single most important variable predicting outcome ($p < 0.001$). Survival rates were 44.7% for hepatitis A, 34.4% for acetaminophen, 23.3% for hepatitis B, 13.6% for drug reaction and 9.0% for non-A, non-B hepatitis. Age was the second most important variable ($p < 0.02$). The final "static" variable found to be independently significant in the overall analysis was the grade of encephalopathy on admission ($p < 0.05$). Among 19 "dynamic" variables analyzed, only serum pH, prothrombin time, admission serum creatinine and bilirubin were found to have statistically significant predictive values. Because of the above findings, patients were then divided into two groups based on underlying etiology of acetaminophen vs. nonacetaminophen-induced liver failure. Table 6.1 summarizes the King's College criteria which predict a poor outcome in acute liver failure patients.

O'Grady and his associates have compiled the best study presently available to determine the optimal timing for transplantation.[15] The major strength of the study is that it bases patient assessment on a combination of easily obtainable predictive markers to establish the risk of death from FHF. In most instances, using these criteria, one can predict the prognosis within a few hours of admission to an emergency room. This will facilitate early referral of those patients with poor prognosis to a specialized liver unit for evaluation for emergency liver transplantation. Its weakness is that the need for transplantation is considered as "nonsurvival" in the context of the study, and hence there is no correlation between the "prognostic" indicators and surgical outcome. The King's College criteria have been validated in a prospective fashion in several series and have since been considered as the standard for predicting outcome based on optimal medical treatment.[16-18]

In a similar study, Bernuau reported an analysis of prognostic criteria in 115 patients with FHF secondary to acute hepatitis B infection.[19] The multivariant analysis revealed that factor V level, patient's age, absence of detectable HbsAg and serum a-fetoprotein concentration were all independent predictors of survival. These criteria were adopted by Bismuth et al and were used to predict outcome in 139 patients with fulminant and subfulminant liver failure admitted to the liver unit at Paul Brousse Hospital in Paris during the period of 1986-1991.[20] Their criteria for liver transplantation were: presence of hepatic encephalopathy (stage 3 or 4) associated with either a factor V level <20% of normal in patients less than 30 years of age, or with a factor V level <30% of normal in patients older than 30 years of age. Among 139 patients who met those criteria on admission, one patient recovered spontaneously, 116 patients underwent orthotopic liver transplantation with a mean waiting time from listing to transplantation of 1.4 days. Twenty-two patients died while awaiting transplantation, with a mean time from admission to death of 1.3 days. Overall survival of patients who underwent transplantation was 62.9%, with a mean follow-up of 32 months

Table 6.1. King's College Hospital criteria for liver transplantation for fulminant hepatic failure

Acetaminophen
pH < 7.30 (irrespective of grade of encephalopathy)
OR all of the following:
Prothrombin time > 100 sec (INR > 6.5)
Serum creatinine > 3.4 g/dl
Grade III or IV hepatic encephalopathy
Nonacetaminophen
Prothrombin time > 100 sec (INR > 6.5) (irrespective of grade of encephalopathy)
OR any 3 of the following variables (irrespective of grade of encephalopathy encephalopathy):
Age < 10 or > 40 yrs
Etiology: Non-A, non-B hepatitis, halothane hepatitis, drug toxicity
Duration of jaundice to encephalopathy > 7 days
Prothrombin time > 50 sec (INR > 3.5)
Serum bilirubin > 17.6 g/dl

and the actuarial 1 and 3 year survival rate was 68.1% and 61.8% respectively. Examining the predictive value of coagulation factors, the King's College group reported their experience with factor V and factor VIII/V ratio on admission. The authors concluded that the predictive value of plasma factor V was lower than that of the King's College criteria in patients with acetaminophen-induced liver failure; however, it may be useful in patients with FHF due to other causes.[21] Other studies have shown that a factor V level less than 10% and a factor VIII/V ratio greater than 30 may enhance predictive accuracy of outcome in FHF secondary to acetaminophen toxicity.[22]

In addition to biochemical and synthetic activities, investigators have considered assessment of the residual functional reserve of the liver as an indicator of prognosis. The ratio of acetoacetate to b-hydroxybutyrate in an arterial blood sample (arterial ketone body ratio—AKBR) is thought to reflect hepatic energy charge. The predictive value of AKBR has been validated in two small series; however, further studies are warranted prior to wider application.[23,24] Another test of functional liver mass is measurement of galactose elimination capacity.[25] Galactose clearance reflects both residual liver mass and hepatic blood flow. This test has been considered as a standard test of liver functional reserve in the past and newer tests are often compared to it.

Van Theil suggested that direct examination of the liver either by imaging or liver biopsy might be a more reliable predictor of prognosis.[26] A liver volume less than 700 cc on computed tomography scan, or more than 50% necrosis on a liver biopsy specimen, served as an indication for transplantation. A comparison between the findings of transjugular liver biopsy and the King's College criteria was reported by Donaldson et al.[18] The King's College criteria accurately predicted outcome in all patients with acetaminophen-induced liver failure and 91% of nonsurvivors and 82% of survivors with nonacetaminophen-induced liver failure. The extent of hepatocyte necrosis on liver biopsy was relatively inaccurate at predicting outcome; however, it correctly predicted outcome in 6 of 8 nonsurvivors and 2 survivors with nonacetaminophen-induced FHF whose composite clinical index was inconclusive by the King's College criteria. Although these criteria seem reasonable, there are a lot of logistical problems associated with both studies and carry a substantial risk for complications, especially in the presence of elevated intracranial pressure (ICP), hemodynamic instability and profound coagulopathy.

Serum Gc-globulin is a plasma protein that is synthesized in the liver. The major function of Gc-globulin is to clear actin released from dying or dead hepatocytes from the extracellular space. It has a short half life, approximately 18 hours for the native protein and less than 30 minutes when bound to actin. Studies in animal models and patients with acute liver failure suggest a profound decrease in serum Gc levels and in the percentage of circulating Gc complexed with actin.[27-29] The predictive accuracy of serum Gc level has been compared to the King's College criteria in several studies. In a study by Tygstrup, Gc levels less than 100 mg/l correctly predicted nonsurvival in 100% of 18 patients with acetaminophen-induced FHF and 79% of 59 patients with nonacetaminophen-induced FHF.[30] In a similar study, 47 consecutive patients with FHF were evaluated to determine whether the admission level of Gc protein could be used to predict survival.[31] The King's College criteria were found to be better than plasma Gc level in predicting the outcome at the time of admission (79% vs. 68%); however, Gc concentration was more accurate after a few days in the hospital (89% vs. 74%). Despite these results, serum Gc-protein assay is presently not widely used and it is not clear whether it will play a major role in predicting prognosis in FHF patients.

In summary, several schemes and prognostic criteria have been suggested and have been used in the clinical setting. Despite this wide variety of indicators, the King's College criteria remain the most widely used for prediction of outcome in FHF patients. These criteria have been validated in several large series and, in most cases, the prognosis could be predicted within a few hours of admission to a medical facility.

Transplant Evaluation

Patients with FHF are admitted to a surgical intensive care unit for management, as discussed in a previous section of the monograph. In addition to aggressive ICU care, transplant surgeons and hepatologists have to assess the patient's overall condition, attempt to determine the underlying etiology of the disease, predict the chances of spontaneous recovery and complete an emergency evaluation for liver transplantation. At our center, the King's College criteria are used as a guideline to predict outcome without liver transplantation. In addition, we follow the general trend in the severity of encephalopathy and elevation of ICP, the degree of coagulopathy, metabolic acidosis and renal failure. Once the initial assessment is made, an emergency evaluation for liver transplantation is completed, usually within 12-24 hours. This evaluation is similar to evaluation of patients with chronic liver disease, with a few exceptions. Patients with FHF usually are not known to have underlying liver disease and the evaluation work-up must reveal the cause for liver failure. As discussed earlier, the differential diagnosis includes:

1. Drug induced hepatotoxicity;
2. Viral hepatitis;
3. Exposure to environmental toxins;
4. Miscellaneous causes; and
5. Indeterminate (cryptogenic) etiology.

In the United States, the most common etiology of FHF remains indeterminate, or non-A, non-B viral hepatitis, which constitutes 45% of all cases in several combined series.[32] Table 6.2 summarizes the laboratory and diagnostic work-up of patients with FHF. In addition to this work-up, patients are seen by various consultants to assess their overall condition and to rule out any absolute contraindication to liver transplantation, as summarized in Table 6.3.

Unlike chronic liver disease, cerebral edema and intracranial hypertension develop rapidly and represent the leading cause of death in this patient population. Neurologic complications resulting in irreversible brain damage or brain-stem herniation are the leading contraindications to liver transplantation in this setting.[33] The cornerstone of patient management in FHF is prevention, early recognition and prompt treatment of cerebral edema. Cerebral edema occurs in later stages of hepatic encephalopathy related to FHF and can be determined by clinical, radiological or invasive methods. Clinical findings include decerebrate posturing, myoclonus, spastic rigidity, seizure activity, systemic hypertension, bradycardia, hyperventilation and mydriasis with diminished pupillary response. Initially, the findings of cerebral edema are paroxysmal, but later on become persistent. Papilledema is a late finding and commonly not present, even in the advanced stages of the disease.[34] Noninvasive diagnostic modalities, including computerized tomographic scanning, EEG monitoring and transcranial doppler flow measurement, have not been proven helpful in early detection or management of cerebral edema.[35-38] Cerebral edema in infants may be diagnosed by cranial ultrasonography to detect lateral ventricle effacement. In older children and adults, CAT scanning can detect cerebral edema, but there is a latent period between its actual onset and radiographic appearance.[36] Magnetic resonance angiography (MRA) is logistically not readily available to critically ill individuals.

Currently, intracranial pressure (ICP) monitoring is the best means of monitoring intracranial hypertension and is recommended for patients with stage 3 or 4 encephalopathy to guide treatment. ICP can be measured using an epidural, subdural or intraventricular fiberoptic catheters. In a large comparative study, data from 262 patients with FHF were

Table 6.2. Liver transplant evaluation for FHF and SFHF

Laboratory Workup	
	CBC and Differential
	Chemistry panel
	Coagulation profile
	24-h Cr Clearance
	Urinalysis
	ABGs
	ANA, AMA, ceruloplasmin, urine copper, a1-antitrypsin
	Tumor markers
	RPR
	Thyroid function tests
	Alcohol and drug toxicology screen
Viral Serologies	
	HAV (IgM, IgG)
	HBV (Hbs Ag, HBV e Ag, HBV DNA)
	HCV
	CMV
	EBV
	HIV
Cultures	
	Bacterial, fungal and viral cultures
	Blood
	Sputum
	Urine
	Ascites
12 lead EKG	
CXR	
Pulmonary function tests	
Abdominal ultrasound with doppler	
Head CT scan	

Table 6.3. Contraindications to liver transplantation in FHF

Absolute Contraindications	
	Severe irreversible brain damage
	HIV infection
	Extrahepatic malignancy
	Uncontrolled sepsis
	Severe pulmonary hypertension and advanced cardiopulmonary disease
	Active substance abuse or major psychosocial problems
Relative Contraindications	
	Elevated ICP or reduced CPP
	Multiorgan system failure
	Hemodynamic instability
	Advanced age
	Portal vein thrombosis except when secondary to hepatocellular carcinoma

analyzed for incidence of bleeding or other complications. Although slightly less sensitive to ICP changes, epidural catheters had the lowest complication rate (3.8%) and lowest rate of fatal hemorrhage (1%).[39] The goal of invasive monitoring is to maintain ICP at levels below 15 mmHg, if possible, while maintaining a cerebral perfusion pressure (CPP) above 50 mmHg. CPP is calculated using the formula: CPP = MAP - ICP; MAP stands for calculated mean arterial pressure. ICP monitoring allows aggressive management of cerebral edema; the same is true for measurement of CPP, which appears to be a better predictor of outcome than ICP alone.[40] Experience with ICP monitoring using a Ladd monitor at the University of Nebraska Medical Center was summarized by Inagaki et al.[41] Data from 61 patients were analyzed in a two part study, initially retrospectively (n = 30) and then prospectively (n = 31); CPP less than 40 mmHg for 2 hours or more was a contraindication to OLT due to the high incidence of irreversible brain damage. Another study, however, of 37 patients with FHF (including 14 patients with adverse parameters, i.e., ICP > 25 mmHg and CPP < 40 mmHg), showed no significant differences in peak ICP, number of ICP surges or lowest CPP among survivors, nonsurvivors and survivors with transplantation.[42] In the same series, 6 out of 14 patients with adverse parameters survived, 2 following liver transplantation. The effect of high ICP, or low CPP, on outcome following liver transplantation, has not been studied in a randomized, controlled fashion. However, it appears that elevated ICP secondary to severe cerebral edema is associated with increased risk of development of irreversible brain damage and a grave outcome following liver transplantation. Therefore, patients with grade III or IV encephalopathy require aggressive and attentive support to prevent development of massive cerebral edema, as delineated in previous sections. Accurate implementation of supportive measures can delay onset of massive cerebral edema. Patients who fail to respond could be placed in a barbiturate coma, as discussed earlier. Thiopental infusion has been shown to decrease cerebral metabolic activity, lower central nervous system oxygen demand and protect the brain from ischemic injury secondary to decreased cerebral blood flow. Thiopental infusion lowered ICP and reduced mortality from FHF in a retrospective, nonrandomized study.[43] In general, however, we find the effect of Thiopental infusion on ICP transient and unpredictable.

Anesthetic Considerations

The anesthetic management of FHF is an extension of the critical care administered in the preoperative period. The anesthesiologist is frequently faced with a patient who may be developing multisystem organ failure and must undergo further physiological stress associated with moving to the operating room, the administration of anesthetic agents and the trauma of surgery.

Measurement of ICP and CPP is critical for the perioperative anesthetic management of these patients. The aim is to maintain the ICP at < 20 mmHg and CPP > 50 mmHg. Mean ICP tends to be elevated more during all the stages of orthotopic liver transplantation than in the preoperative period. Indeed, all but one patient in Lidofsky's series had transient intracranial hypertension requiring medical intervention at some stage of the procedure regardless of whether they had it preoperatively.[35] Intraoperative ICP changes among [14] patients undergoing OLT for FHF were reviewed retrospectively at our institution.[44] There was a statistically significant increase in mean ICP during the hepatectomy phase (14.8 ± 4.1 to 21.0 ± 5.4 mmHg, $p = 0.02$). This was followed by a significant lowering of the ICP after removal of the necrotic liver mass (9.7 ± 2.2 mmHg), followed by a second peak of ICP elevation with graft reperfusion (15.5 ± 2.0 mmHg). These ICP changes are multifactorial, most likely due to adoption of horizontal position on the operating table, administration of anesthetic agents, noxious stimuli of surgery and to progressive fluid and electrolyte changes associated with the operation. Transient increases in ICP may occur at any stage of the procedure, are not always associated with hemodynamic changes and should be treated aggressively. Elevation of the head, deepening of anesthesia, adequate neuromuscular blockade, mannitol (with due consideration to renal function, serum osmolarity) and/or barbiturates (pentobarbitone 3-5 mg/kg i.v. followed by an infusion of 1-3 mg/kg/h) could be used. It was originally postulated that veno-venous bypass would protect the brain during OLT; however, few centers have shown the feasibility of doing the procedure without bypass.[10,45]

The anhepatic phase of the operation is frequently the safest with regard to ICP elevation. This supports performance of early total hepatectomy in the severely ill patient with very high ICP and impending brain stem herniation even if an organ is not immediately available for transplantation.[45] Lidofsky reported that 50% of FHF patients undergoing OLT demonstrated a gradual increase in ICP during this phase and that pressure should be maintained with inotropic agents if necessary to improve cerebral perfusion.[45] Phenylephrine, epinephrine, dopamine, dobutamine and calcium chloride have all been used successfully. Severe systemic hypertension is best treated with adrenergic blocking agents. Prager reported that Isoflurane tended to increase ICP in 6 of 7 patients.[46]

Fluid replacement and electrolyte correction may be very difficult to manage, particularly in the face of renal failure. The use of blood products should be monitored closely. Intraoperatively, coagulation should be monitored by the thromboelastograph.[47]

Surgical Technique

There are currently several techniques available for liver transplantation, some of which apply only to patients with FHF. Generally, liver transplantation in this group of patients can be done with an orthotopic or a heterotopic approach. Orthotopic liver transplantation involves a total hepatectomy followed by placement of an allograft in the anatomic position. In heterotopic liver transplantation (HLT), the native liver is left intact and the liver allograft is placed in a nonanatomic position. Orthotopic liver transplantation has several advantages that include removal of a necrotic liver, adequate space for placing the new liver, ease of anatomic alignment, a single organ to assess for function and use of commonly applied transplantation techniques. The standard techniques of orthotopic liver transplantation, developed by Starzl and his associates and modified by others, are well described in the surgical literature and will not be further elaborated in this chapter.[48-51] There are, however, several negative aspects of OLT relative to HLT in the setting of FHF. They include greater risk of bleeding during the hepatectomy phase because many of these patients have profound coagulopathy, removal of an organ which (if not completely necrotic) may regenerate and become functional again, need for life-long immunosuppression and increased risk of developing malignancies. Despite these potential disadvantages, OLT still is the most commonly used approach for transplanting FHF patients.

In the classic orthotopic approach, the infra- and supra-hepatic vena cava are clamped, the patient is placed on portosystemic veno-venous bypass and the retro-hepatic IVC is excised with the liver.[52] Initially, veno-venous bypass was thought to be essential for maintaining hemodynamic stability during the anhepatic phase; however, several studies have shown the feasibility of OLT without veno-venous bypass.[52,53] Using the so-called "piggyback" technique, total hepatectomy is completed with preservation of the retro-hepatic IVC, thus obviating the need for clamping the IVC.[10,54,55] The donor supra-hepatic IVC is sewn to a caval cuff fashioned by the tributaries of the left, middle and possibly right hepatic veins, while the infra-hepatic IVC stump is oversewn (Fig. 6.1). The advantage of this procedure is to avoid the hemodynamic changes associated with clamping the IVC and veno-venous bypass; in addition, it involves only a single caval anastomosis. This technique

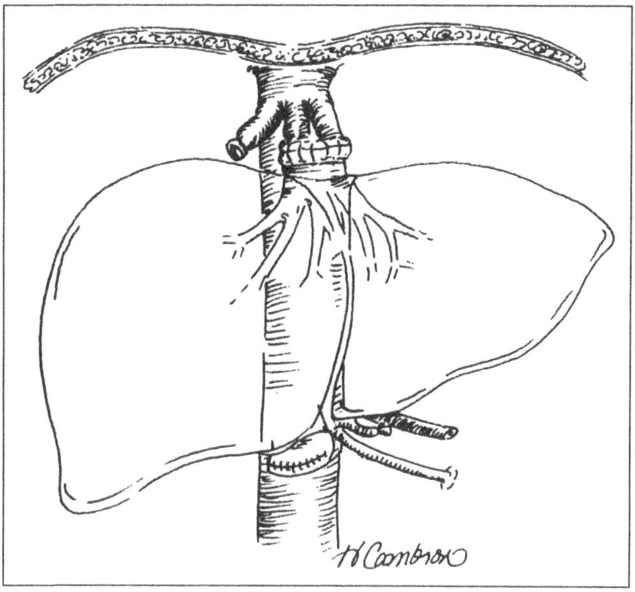

Fig. 6.1. Orthotopic liver transplantation using the "piggy back" technique. 1. Total hepatectomy with preservation of the retrohepatic IVC; 2. Implantation of the allograft with a single suprahepatic caval anastomosis. Notice that the donor infrahepatic IVC stump has been oversewn.

has been used for patients with FHF; however, it is not performed routinely in this setting due to technical considerations.

The results of liver transplantation for patients with FHF using the classic technique have been variable. Ascher et al reported 1 year actuarial survival rates of 92% and 100% in 35 patients with FHF and SFHF respectively.[10] On the other hand, Bismuth et al reported a 68% 1 year survival following OLT in 116 patients with FHF.[20] The primary reason for this difference in survival is the difference in patient selection criteria among various centers. Exclusion of high-risk patients (patients with high ICP and low CPP) improves OLT survival. A more meaningful comparison of center-specific survival data can be carried out if the survival of the total number of FHF patients treated at a center is reported, not just survival of patients undergoing OLT.

HLT was first attempted in the laboratory by Welch and coworkers in 1955.[56] More recently, the procedure was popularized by Terpstra.[57,58] In HLT, the allograft is placed in the right paravertebral gutter after mobilizing the right colon and the hepatic flexure medially. Vascular anastomoses are performed in a nonanatomic fashion in contrast to the orthotopic procedure, commonly using an end to side configuration. The supra-hepatic IVC of the donor is anastomosed in an end to side fashion to the recipient's infra-renal vena cava, while the donor infra-hepatic vena cava is oversewn. The donor portal vein is anastomosed to the recipient's portal vein in an end to side fashion. The donor hepatic artery with a Carrel patch of the aorta is sutured to the recipient's infrarenal aorta. Finally, biliary reconstruction is completed with a Roux-en-Y hepatico-jejunostomy. The alignment of the native blood vessels and bile duct is not disturbed. A major disadvantage of HLT is the lack of adequate space in the abdominal cavity; thus, selection of either small or reduced size grafts may be necessary. Unlike patients with chronic liver disease, long standing tense ascites does not usually occur in FHF and the abdominal wall is usually not stretched. If the graft is too large, abdominal closure may be difficult and may result in compression of the venous structures, or it can cause respiratory failure and need for prolonged intubation.

According to Shaw, auxiliary HLT is an ideal treatment for FHF if the following criteria are met:

1. Recovery of the native liver is a certainty;
2. The allograft fits into an adequate heterotopic position;
3. The venous outflow and portal venous inflow to the graft are satisfactory for the duration of required support.[59]

Moritz was the first to report successful use of HLT in a 19-year-old woman with non-A, non-B, non-C hepatitis with FHF and stage IV hepatic encephalopathy; the patient recovered fully. He and his associates chose the heterotopic position, citing an adverse effect of manipulation of the necrotic liver on ICP, causing severe hemodynamic instability. Despite the use of a reduced size donor liver, which was placed in the right paravertebral gutter, abdominal wound closure required a silastic mesh. The surgeons reported that native liver biopsy at 6 months showed cirrhosis; however, subsequent evaluation at 22 months revealed normal native liver histology and absent hepatocytes with diffuse fibrovascular (granulation) tissue in the graft. Imaging studies revealed a shrunken graft with no hepatocyte function and a native liver of normal size and function. Immunosuppression was withdrawn and the graft allowed to undergo atrophy. At 44 months following HLT, the patient continued to have normal life and normal liver function.[61] Several other cases have been described since the initial report; however, the experience with HLT remains limited.[62-64] Most of the reported graft failure in HLT is due to technical reasons, primarily inadequate portal venous flow due to portal "steal", with preferential blood flow into the native portal vein or compression of either the portal vein or IVC due to inadequate space. At this time, there have been no reports of removal of the native liver and long-term survival with an allograft in the heterotopic position.

Auxiliary, partial, orthotopic liver transplantation (APOLT) was developed as a result of the increasing interest in "bridging" patients until the native liver regenerates. The orthotopic position has two advantages over the heterotopic position. First, the graft hepatic vein can be implanted into an anatomic position and provide better venous drainage of the liver. Second, it reduces the volume of both native and donor livers, thus facilitating abdominal closure. Tight abdominal closure jeopardizes portal flow into the native liver by increasing parenchymal resistance, already elevated due to massive necrosis; and, subsequently, it affects the ability of the native liver to regenerate. Boudjema reported the first series of APOLT in FHF and SFHF. The procedure includes:

1. Partial native liver resection (left lateral segment, left lobe, or right lobe);
2. Back-table segmental resection (Couinaud's classification) of the donor liver;
3. Implantation of the reduced size graft into its orthotopic position beside the remaining native liver (Fig. 6.2).

Four of 8 patients (50%) demonstrated complete regeneration of the native liver within 4 months, immunosuppression was discontinued and the allograft was surgically removed.[65] Similar results were reported by the European multicenter study summarizing the experience with 30 patients who underwent APOLT for FHF.[63] After a median follow-up of 11 months (range 3-67 mo), 63% of the patients were alive and 13 (43%) recovered normal native liver function. The percentage and distribution of necrosis observed on intraoperative biopsy of the native liver were not related to the final outcome. More important factors were the age of a patient and underlying disease etiology. Bismuth et al reported complete native liver regeneration in 3 of 5 patients who underwent AOPLT for FHF secondary to Reye's syndrome, HAV and HBV.[66] All 3 patients were under the age of 40 years. Additional experience

Fig. 6.2. Auxiliary partial orthotopic liver transplant (APOLT). Left hepatic allograft is implanted in the orthotopic position following a recipient left hepatic lobectomy. Both the vena cava and portal vein are kept in their anatomic position with end to end anastomosis to the recipient vessels.

with AOPLT was reported recently by Shaw et al showing native liver regeneration in 6 of 7 patients with severe liver injury and hepatocyte necrosis between 50-100%.[67] Three patients were alive with a nonfunctioning allograft after withdrawal of immunosuppression, while in the remaining four the grafts had to be surgically removed.

Living related liver transplantation (LRLT), either auxiliary or total, has become an accepted surgical modality for treatment of children with chronic liver disease due to the shortage of pediatric organ donors; however, only a few cases have been reported in the setting of FHF.[68-70] The critical issue in LRLT is the adequacy of the size of the graft harvested from the donor. Extensive donor hepatectomy may jeopardize donor safety, while a small graft may be inadequate to handle the metabolic needs of the recipient. The use of a right lobe provides a larger graft; however, harvesting a right lobe for LRLT is technically more difficult and can result in increased operative morbidity and mortality. The left lateral segment, or left lobe, is the preferred graft to use and the donor operation is safer.[71] The minimal effective graft size remains unknown; however, limited data from the LRLT experience suggest that a minimum graft weight to recipient body weight ratio of 0.5-1.9%, or a graft volume to recipient estimated liver volume ratio of 25%, is necessary for immediate graft function.[69,72,73]

With the increasing shortage of organ donors, few patients with FHF may develop a "toxic liver syndrome" characterized by severe intracranial hypertension, profound lactic acidosis, hemodynamic instability and multisystem organ failure. Ringe et al reported their experience in patients with primary allograft nonfunction who experienced immediate reversal of refractory shock following allograft removal.[74] This observation was supported by Husberg, who reported three patients with FHF and severe lactic acidosis which was corrected by hepatic devascularization.[75] These and other clinical observations led to the suggestion that removal of the necrotic liver may result in improved hemodynamic status and a decrease in ICP. In such extreme cases, a two-stage procedure is applied: total hepatectomy with end to side portocaval shunt followed by liver transplantation when an allograft becomes available. The largest series of patients with FHF rendered anhepatic while awaiting liver transplantation was recently reported by Ringe et al.[76] Thirty-two adults with "toxic liver syndrome" underwent total hepatectomy and portocaval shunt. Thirteen patients did not show any signs of improvement after hepatectomy and died rapidly from multisystem organ failure, while 19 became more stable and underwent the full procedure. Patients were anhepatic for 987 ± 433 min (range 395-2489 min). Only seven patients remained alive at follow-up of 3-46 months.

We have used this approach to treat an 18-year-old female patient with uncontrollable cerebral edema secondary to FHF; she underwent total hepatectomy and portocaval shunt followed by OLT 14 hours later.[77] During the anhepatic period she was treated with an extracorporeal liver support system (bioartificial liver) as described in another section of this monograph. With artificial liver support, there was reversal of severe neurological dysfunction, normalization of intracranial pressure and a decrease in serum ammonia. The patient had a complete recovery without any neurologic deficit. A second patient with FHF has also been treated with the same approach at our Unit with a successful long term outcome.

It appears that in highly selected patients with severe "toxic metabolic state" and uncontrollable intracranial hypertension, total hepatectomy and portocaval shunt, preferably with some form of artificial liver support, followed by OLT may be considered as a desperate measure to salvage these patients.

Results

Most major series presenting the experience with OLT in FHF since the mid-1980s report 58-92% patient survival; however, not all present true or actuarial one year patient survival data. The orthotopic technique has become the procedure of choice due to better survival and probably greater familiarity by the surgeons. Similarly, transplantation of pediatric patients has also been successful. In a recent study of 35 children between ages 3 months to 15 years, there was a 68% survival at 6 months to 4 years of follow-up using the orthotopic technique.[78] Compared to the 20% overall survival of patients with FHF following medical management alone, one can easily see the major impact liver transplantation has had on the treatment of this disease. The improvement in survival is even more dramatic if one excludes patients with acetaminophen toxicity. Based on these results, liver transplantation should be offered as a therapeutic option to potentially all patients with severe FHF if they meet appropriate criteria (poor prognostic indicators) as described above.

In general, postoperative management is similar to that of patients undergoing liver transplantation for chronic liver disease. Aside from routine ICU management, specific treatment of certain complications commonly associated with increased mortality in the setting of transplantation for FHF is needed. These complications include severe neurologic dysfunction related to residual cerebral edema, increased risk of infection, multisystem failure, graft rejection due to ABO incompatibility and graft loss. The most effective method of treatment for the first two problems is institution of careful selection criteria and aggressive preoperative measures, as delineated earlier in this chapter. Devastating neurologic complications are encountered less frequently as clinicians become more experienced and relative exclusion criteria become better defined.[6,10,79] The spectrum of neurologic complications ranges from transient aphasia to brain death.[5,7,9,79] Continued use of ICP monitoring is necessary to guide therapy until ICP normalizes and adequate hepatic function is established. Removal of the ICP monitor should not be performed before correcting significant coagulopathy, due to the increased risk of intracranial hemorrhage or hematoma formation.

Occult sepsis is commonly seen in FHF.[80] Several series report a significant incidence of bacterial and fungal sepsis.[3,6,8] Infection is most commonly caused by gram-positive organisms, predominantly *Staphylococcus aureus*. Gram-negative organisms, especially coliforms, are the cause for at least 30% of the infections. *Candida* and *Aspergillus* are the most common fungal pathogens and constitute up to 20% of the infections in FHF.[81,82] Broad-spectrum antibiotic and antifungal therapy is warranted, as well as close clinical observation and routine surveillance cultures. Antibiotics are discontinued 48 hours following transplantation unless warranted from culture results or clinical suspicion, to lower the risk of fungal super-infection. Due to the high incidence of fungal infection and its associated high morbidity and mortality, prophylactic use of intravenous fluconazole or amphotericin B is recommended. In the presence of renal dysfunction, liposomal amphotericin B may be used.

Cytomegalovirus (CMV) infection is the most commonly described viral infection following transplantation for FHF.[2,5] The range of clinical presentation varies from asymptomatic to invasive disease presenting with fever, pneumonia, hepatitis, gastrointestinal ulcers, enteritis or colitis.

Recurrence of viral hepatitis following transplantation for FHF, is an important issue.[83] Recurrence of hepatitis B is well documented. Interestingly, in contrast to patients transplanted with chronic hepatitis B, recurrent hepatitis and hepatic dysfunction are less common following transplantation for the fulminant form of the disease.[84,85] Samuel and his coworkers reported HBsAg-negativity in 17 patients transplanted for FHF secondary to hepatitis B following passive immunoprophylaxis with hepatitis B immunoglobulin (HBIg).[85] Lifetime administration of HBIg may be necessary to prevent recurrence. In addition to HBIg, prophylaxis with an antiviral agent (lamivudine) has been recently used, alone or in combination with HBIg, in patients with fulminant HBV infection; however, there are no data to support wide application of this practice at this time. Unlike viral hepatitis B, FHF secondary to hepatitis C is uncommon. Several US and European series suggest a low incidence of hepatitis C resulting in FHF based upon PCR determinations.[86-89] Because hepatitis C RNA is present only rarely in indeterminate cases of FHF, it is difficult to ascribe to hepatitis C a significant causality of FHF.

The morbidity associated with acute cellular rejection following transplantation for FHF is difficult to assess from examining the literature. One series demonstrates absence of acute rejection, while others suggest a high incidence associated with high morbidity.[5,9,90] Early graft dysfunction is frequently, and probably correctly, attributed to poor donor selection; however, antibody-mediated (humoral) accelerated graft rejection can be present in certain cases and is very difficult to diagnose. A specific immunosuppressive regimen following transplantation cannot be recommended based on the existing information. Judicious usage of antirejection agents so that optimal serum levels and graft function are present and concurrent administration of appropriate bacterial/viral/fungal prophylaxis or treatment are needed. With the availability of FK 506 and mycophenolate, induction immunosuppressive therapy with antithymocyte globulin (ATG) or OKT-3 is no longer used in patients with renal failure.[91]

Due to the rapid progression of the disease process in FHF patients and the urgent need for hepatectomy and transplantation, suboptimal donor allografts may be considered, as timely retrieval of a satisfactory organ may not be possible.[3,7] Primary graft nonfunction is probably more common following transplantation for FHF than for other etiologies of liver disease and is associated with high morbidity and mortality.[5] Primary nonfunction could be associated with cerebral edema and underlying sepsis.[92,93] Hepatic dysfunction can result in progressive cerebral edema or poor clearance of infectious pathogens due to inadequate reticuloendothelial function.

Use of ABO-incompatible grafts has been effective in the treatment of FHF in adults and children.[3,6,7,10,78,94-97] ABO-incompatible graft can achieve permanence or, more likely, it will act as a "bridge", at least until the cerebral edema has dissipated. Upon failure of the incompatible graft, retransplantation with an ABO-compatible graft is performed in a more stable patient with less concern for neurologic injury. Though the shortage of donor organs must be considered, not offering transplantation may result in the loss of an otherwise salvageable patient.

Additional short and long term complications following liver transplantation for FHF or SFHF are similar to those encountered in patients undergoing liver transplantation for chronic liver disease. These complications have been described elsewhere and include biliary complications in the form of leak or stricture, vascular complications such as hepatic artery or portal vein thrombosis or vena cava stenosis. Other complications include infection, chronic rejection, recurrence of the primary disease and increased risk for posttransplant lymphoproliferative disorder and other malignancies due to long term immunosuppression.[92]

Summary

FHF is a devastating condition affecting approximately 5000 patients annually. Presently, liver transplantation is the only effective treatment modality resulting in excellent long term survival. The severe organ shortage necessitates a relatively long waiting time during which patients may develop irreversible brain damage. Spontaneous regeneration of the native liver with aggressive medical management is known to occur in at least 20% of FHF patients; however, there are currently no known prognostic criteria allowing accurate and reliable (100%) prediction of which patient will recover spontaneously and which will require a transplant.

OLT is the most commonly used surgical approach for transplanting patients with FHF. Other techniques have been described and are being routinely used in a few transplant centers. Auxiliary liver transplantation in carefully selected patients offers the advantage of "bridging" patients until their native livers regenerate, with subsequent withdrawal of immunosuppression and atrophy of the allografts. Several transplant centers have reported complete functional and histologic regeneration of the native liver following documented initial massive necrosis. Young patients with FHF secondary to acetaminophen toxicity, acute HAV or HBV should be considered for partial auxiliary orthotopic liver transplantation in centers which have the surgical expertise. Finally, to achieve good outcomes in FHF, early diagnosis and prompt patient referral to specialized centers is needed.

Acknowledgment

The authors would like to thank Ms. Helen Cambron, M.S.N., FNP for providing the medical illustration and figures.

References

1. Tygstrup N, Ranek L. Assessment of prognosis in fulminant hepatic failure. Semin Liver Dis 1986; 6:129-37.
2. Schafer DF, Shaw BW. Fulminant hepatic failure and orthotopic liver transplantation. Semin Liver Dis 1989; 9:189-94.
3. Emond JC, Aran PP, Whitington PF et al. Liver transplantation in the management of fulminant hepatic failure. Gastroenterology 1989; 96:1583-88.

4. Peleman RP, Galaver JS, Van Thiel DH et al. Orthotopic liver transplantation for acute and subacute hepatic failure in adults. Hepatology 1987; 7:484-89.
5. Campbell DA Jr, Ham JM, McCurry KR et al. Liver transplantation for fulminant hepatic failure. American Surgeon 1991; 8:546-49.
6. Isai H, Heil AGR, McCaughan GW et al. Fulminant hepatic failure and liver transplantation. Transplantation Proc 1992; 24:1475.
7. Bismuth H, Samuel D, Gugenheim J et al. Emergency liver transplantation for fulminant hepatitis. Ann Int Med 1987; 107:337-41.
8. O'Grady JG, Alexander GJM, Thick M et al. Outcome of orthotopic liver transplantation in the aetiological and clinical variants of acute liver failure. Q Jr Med 1988; 258:817-24.
9. Brems JJ, Hiatt JR, Ramming KP et al. Fulminant hepatic failure: The role of liver transplantation as primary therapy. Am J Surg 1987; 154:137-41.
10. Ascher NL, Lake JR, Emond JC et al. Liver transplantation for fulminant hepatic failure. Arch Surg 1993; 128:677-82.
11. Belle SH, Beringer KC, Detre KM. Recent findings concerning liver transplantation in the United States. In: Cecka JM, Terasaki PI, eds. Clinical Transplants. 12 ed. Los Angeles: UCLA Tissue Typing Laboratory, 1996:15-29.
12. Trey C, Davidson C. The management of fulminant hepatic failure. Prog Liver Dis 1970; 3:282-98.
13. Bernuau J, Rueff B, Benhamou J-P. Fulminant and subfulminant liver failure: Definitions and causes. Semin Liver Dis 1986; 6:97-106.
14. Christensen E, Bremmelgaard A, Bahnsen M et al. Prediction of fatality in fulminant hepatic failure. Scand Jr Gastroenterology 1984; 1990-96.
15. O'Grady JG, Alexander GJM, Hayallar KM et al. Early indicators of prognosis in fulminant hepatic failure. Gastroenterology 1989; 97:439-45.
16. Lee WM, Galbraith RM, Watt GH et al. Predicting survival in fulminant hepatic failure using serum Gc protein concentration. Hepatology 1995; 21:101-105.
17. Pauwels A, Mostefa-Kara N, Flornet C et al. Emergency liver transplantation for acute liver failure: Evaluation of London and Clichy criteria. Hepatology 1993; 17:124-127.
18. Donaldson BW, Gopinath R, Wanless IR et al. The role of transjugular liver biopsy in FHF: A relation to other prognostic indicators. Hepatology 1993; 18:1370-1374.
19. Bernuau J, Goudean A, Poynard T et al. Multivariate analysis of prognostic factors in fulminant hepatitis B. Hepatology 1986; 6:648-51.
20. Bismuth H, Didier S, Castaing D et al. Orthotopic liver transplantation in fulminant and subfulminant hepatitis: The Paul Brousse experience. Ann Surg 1995; 222:109-119.
21. Izumi S, Langley PG, Wendon J et al. Coagulation factor V levels as a prognostic indicator in fulminant hepatic failure. Hepatology 1996; 23:1507-1511.
22. Pereira LMMB, Langley PG, Hayllar KM et al. Coagulation factor V and VIII/V ratio as predictors of outcome in paracetamol induced fulminant hepatic failure: Relation to other prognostic indicators. Gut 1992; 33:98-102.
23. Saibara T, Onishi S, Sone J et al. Arterial ketone body ratio as a possible indicator for liver transplantation in fulminant hepatic failure. Transplantation 1991; 51:782-786.
24. Scaiola A, MacMathuna P, Langley PG et al. Determination of the ketone body ratio in fulminant hepatic failure. Hepatogastroenterology 1990; 37:413-416.
25. Ranek L, Andreasen PB, Tygstrup N. Galactose elimination capacity as a prognostic index in patients with fulminant hepatic failure. Gut 1976; 17:959-964.
26. Van Thiel DH. When should a decision to proceed with transplantation be made in cases of fulminant or subfulminant hepatic failure: At admission to hospital or when a donor organ is made available? J Hepatology 1993; 17:1-2.
27. Lee WM, Emerson DL, Werner PM et al. Decreased serum group-specific component protein levels and complexes with actin in fulminant hepatic necrosis. Hepatology 1985; 5:271-275.
28. Lee WM, Young WO, Goldschmidt-Clermont PJ et al. Diminished serum Gc (vitamin D-binding protein) levels and increased Gc:G-actin complexes in a hamster model of fulminant hepatic necrosis. Hepatology 1987; 7:825-830.

29. Young WO, Goldschmidt-Clermont PJ, Emerson DL et al. Correlation between extent of liver damage in fulminant hepatic necrosis and complexing of circulating Gc (Vitamin D-binding protein). J lab Clin Med 1987; 110:83-90.
30. Schiodt FV, Bondesen S, Petersen I et al. Admission levels of serum Gc-globulin: Predictive value in fulminant hepatic failure. Hepatology 1996; 23:713-718.
31. Lee WM, Galbraith RM, Watt GH et al. Predicting survival in fulminant hepatic failure using serum Gc protein concentrations. Hepatology 1995; 21:101-105.
32. McCashland TM, Shaw BW, Tape E. The American experience with liver transplantation for acute liver failure. Seminar in Liver Disease 1996; 16:427-433.
33. Lidofsky SD. Fulminant hepatic failure. Crit Care Clin 1995; 11:515-430.
34. Williams R, Gimson AES. Intensive liver care and management of acute hepatic failure. Dig Dis Sci 1991; 36:820-826.
35. Lidofsky SD, Bass NM, Prager MC et al. Intracranial pressure monitoring and liver transplantation for fulminant hepatic failure. Hepatology 1992; 16:1-7.
36. Munoz SJ, Robinson M, Northrup B et al. Elevated intracranial pressure and computed tomography of the brain in fulminant hepatic failure. Hepatology 1991; 13:209-212.
37. Wijdicks EFM, Plevak DJ, Rakela J et al. Clinical and radiologic features of brain edema in fulminant hepatic failure . Mayo Clin Proc 1995; 70:119-124.
38. Sidi A, Mahla ME. Noninvasive monitoring of cerebral perfusion by transcranial doppler during fulminant hepatic failure and liver transplantation. Anesth Analg 1995; 80:194-200.
39. Blei AT, Olafsson S, Webster S et al. Complications of intracranial pressure monitoring in fulminant hepatic failure. Lancet 1993; 341:157-58.
40. Jenkins JG, Glasgow JF, Black GW et al. Reye's syndrome: Assessment of intracranial pressure monitoring. Br Med J 1987; 294:337-338.
41. Inagaki M, Shaw BW Jr, Schafer DF et al. Advantages of intracranial pressure monitoring in patients with fulminant hepatic failure. Gastroenterology 1992; 102:A826.
42. Aldersley MA, Juniper M, Richardson P et al. Inability of intracranial pressure monitoring to select patients with acute liver failure for transplantation. Hepatology 1995; 22:A408.
43. Forbes A, Alexander GJM, O'Grady JG et al. Thiopental infusion in the treatment of intracranial hypertension complicating fulminant hepatic failure. Hepatology 1989; 10:306-310.
44. Lidofsky SD, Bass NM, Prager MC et al. Intracranial pressure monitoring and liver transplantation for fulminant hepatic failure. Hepatology 1992; 16:1-7.
45. Detry O, Kahaku E, Arkadopoulos N et al. Intracranial pressure during liver transplantation for fulminant liver failure. Transplantation 1997, In Press.
46. Keays R, Potter D, O'Grady J et al. Intracranial and cerebral perfusion pressure changes before, during and immediately after orthotopic liver transplantation for fulminant hepatic failure. Q J Med 1991; 79:425- 433
47. Prager MC, Huk I, Kretschmer G et al. Transplant Proceedings 1993; 25:1841.
48. Kang YG, Martin DJ, Marquez J et al. Intraoperative changes in blood coagulation and thromboelastographic monitoring in liver transplantation. Anesth Analg 1985; 64:888-896.
49. Starzl TE, Marchioro TL, Von Kaulla KN et al. Homotransplantation of the liver in humans. Surg Gynecol Obstet 1963; 117:659-676.
50. Starzl TE, Iwatsuki S, Esquivel CO et al. Refinements in the surgical technique of liver transplantation. Semin Liver Dis 1985; 5:385-389.
51. Denmark SW, Shaw BW Jr, Starzl TE et al. Veno-venous bypass without systemic anticoagulation in canine and human liver transplantation. Surg Forum 1983; 34:380-382.
52. Tzakis AG, Todo S, Steiber AC et al. Venous jump grafts for liver transplantation in patients with portal vein thrombosis. Transplantation 1989; 48:530-531.
53. Estrin JA, Belani KG, Ascher NL et al. Hemodynamic changes on clamping and unclamping major vessels during liver transplantation. Transp Proc 1989; 21:3500-3505.
54. Carmichael FJ, Lindop MJ, Farman JV. Anesthesia for hepatic transplantation. Anesth Analg 1985; 64:108-116.
55. Tzakis A, Todo S, Starzl TE. Orthotopic liver transplantation with preservation of the inferior vena cava. Ann Surg 1989; 210:649-652.

56. Fleitas MG, Casanova D Martino E et al. Could the piggyback operation in liver transplantation be routinely used. Arch Surg 1994; 842-845.
57. Welch CS. A note on transplantation of the whole liver in dogs. Transpl Bull 1955; 2:54-59.
58. Terpstra OT, Schalm SW, Reuvers CV et al. Auxiliary heterotopic liver transplantation. Transplantation 1988; 45:1003-1007.
59. Terpstra OT, Schalm SW, Weimar W et al. Auxiliary partial liver transplantation for end stage chronic liver disease. NEJM 1988; 319:1507-1511.
60. Shaw BW. Auxiliary liver transplantation for acute liver failure. Liver Transplantation and Surg 1995; 1:194-200.
61. Moritz MJ, Jarrell BE, Armenti V et al. Heterotopic liver transplantation for fulminant hepatic failure- a bridge to recovery. Transplantation 1990; 50:524-526.
62. Mortiz MJ, Jarrell BE, Munoz SJ et al. Regeneration of the native liver after heterotopic liver transplantation for fulminant hepatic failure. Transplantation1993; 55:952-954
63. Stampf DA, Munoz SJ, Moritz MJ et al. Heterotopic liver transplantation for fulminant Wilson's disease. Gastroenterology 1990; 99:1834-1836.
64. Chenard-Neu MP, Boudjema K, Bernau J et al. Auxiliary liver transplantation: Regeneration of the native liver and outcome in 30 patients with fulminant hepatic failure—A multicenter European study. Hepatology 1996; 23:1119-1127.
65. Metselaar HJ, Hesselink EJ, De Rave S et al. Recovery of failing liver after auxiliary heterotopic transplantation. Lancet 1990; 335:1156.
66. Boudjema K, Cherqui DJ, Jaeck D et al. Auxiliary liver transplantation for fulminant and subfulminant hepatic failure. Transplantation 1995; 59:218-223.
67. Bismuth H, Azouly D, Samuel D et al. Auxiliary partial liver transplantation for fulminant hepatitis. Ann Surg 1996; 6:712-726.
68. Sudan DL, Shaw BW. Long-term follow-up of auxiliary liver transplantation for fulminant hepatic failure. Transplantation Proceedings 1997; 29:485-486.
69. Egawa H, Tanaka K, Inomata Y et al. Auxiliary partial liver transplantation from a living related donor: A report of two cases. Transplantation Proceedings 1996; 28:1071-1072.
70. Fuchinoue W, Tanaka K, Takasaki K et al. Living-related liver transplantation for fulminant hepatic failure. Transplantation Proceedings 1997; 29:424-427.
71. Emond JC, Renz JF, Ferrell LD et al. Functional analysis of grafts from living donors. Implications for the treatment of older recipients. Ann Surg 1996; 224:544-554.
72. Malago M, Rogiers X, Burdelski M et al. Living related liver transplantation: 36 cases at the University of Hamburg. Transplantation Proceedings 1994; 26:3620.
73. Habib N, Tanaka K. Living-related liver transplantation in adult recipient: A hypothesis. Clin Transplantation 1995; 9:835.
74. Chung-Mau L, Sheung-Tat F, Chan JK et al. Minimum graft volume for successful adult-to-adult living donor liver transplantation for fulminant hepatic failure. Transplantation 1996; 62:696-698.
75. Ringe B, Pichlmayr R, Lubbe N et al. Total hepatectomy as a temporary approach to acute hepatic or primary graft failure. Transplantation Proceedings 1988; 20:552-557.
76. Husberg BS, Goldstein RM, Klintmalm GB et al. A totally failing liver may be more harmful than no liver at all: Three cases of total hepatic devascularization in preparation for emergency liver transplantation. Transplantation Proceedings 1991; 23:1533-1535.
77. Ringe B, Lubbe N, Kuse E et al. Total hepatectomy and liver transplantation as a two-stage procedure. Annals Surg. 1993; 218:3-9.
78. Rozga J, Podesta L, LePage E et al. Control of cerebral edema by total hepatectomy and extracorporeal liver support in fulminant hepatic failure. Lancet 1993; 342:898-899.
79. Devictor D, Desplanques L, Debray D et al. Emergency liver transplantation for fulminant liver failure in infants and children. Hepatology 1992; 16:1156-1162.
80. Steiber AC, Ambrosino G, Van Thiel D et al. Orthotopic liver transplantation for fulminant and subacute hepatic failure. Gastroenterology Clinics of North America 1988; 17:157-165.
81. Rolando N, Alexander GJM, Harvey F et al. Prospective study of bacterial infection in acute liver failure: An analysis of 50 patients. Hepatology 1990; 2:49-53.

82. Rolando N, Harvey F, Brahm J et al. Fungal infection: A common, unrecognized complication of acute liver failure. J Hepatol 1991; 12:1-9.
83. Rolando N, Gimson AES, Wade J et al. Prospective controlled trial of selective parenteral and enteral antimicrobial regimen in fulminant liver failure. Hepatology 1993; 17:196-201.
84. Martin P, Munoz SJ, Friedman LS. Liver transplantation for viral hepatitis: current status. Am J Gastroenterology 1992; 4:409-18.
85. Todo S, Demetris AJ, Van Thiel et al. Orthotopic liver transplantation for patients with hepatitis B virus-related liver disease. Hepatology 1991; 13:619-626.
86. Samuel D, Bismuth H, Mathieu D et al. Passive immunoprophylaxis after liver transplantation in HBsAg-positive patients. Lancet 1991; 337:813-815.
87. Feray C, Gigou M, Samuel D et al. Hepatitis C virus RNA and hepatitis B virus DNA in serum and liver of patients with fulminant hepatitis. Gastroenterology 1992; 103:549-555.
88. Liang TJ, Jeffers L, Reddy RK et al. Fulminant or subfulminant non-A, non-B viral hepatitis: Hepatitis C and E viruses. Gastroenterology 1992; 103:556-562.
89. Wright TL, Hsu H, Donegan E et al. Hepatitis C virus not found in fulminant non-A, non-B hepatitis. Ann Int Med 1991; 115:111-12.
90. Sallie R, Silva AE, Purdy M et al. Hepatitis C and E in non-A non-B fulminant hepatic failure: A polymerase chain reaction and serological study. Journal of Hepatology 1994; 20:580-588.
91. Mor E, Solomon H, Gibbs JF et al. Acute cellular rejection following liver transplantation: Clinical pathologic features and effect on outcome. Semin Liv Dis 1992; 12:28-40.
92. Eckhoff DE, McGuire BM, Frenette LR et al. Tacrolimus (FK 506) and mycophenolate mofetil combination therapy versus tacrolimus in adult liver transplantation. Transplantation 1998; 65:180-187.
93. Busuttil RW, Klintmalm GB eds. Transplantation of the liver. WB Saunders Co, Philadelphia, PA 1986.
94. Greig PD, Woolf GM, Sinclair SB et al. Treatment of primary liver graft nonfunction with prostaglandin E1. Transplantation 1989; 48:447-453.
95. Mollison LC, Angus PW, Jones RM et al. Prostaglandin E1 for the treatment of primary nonfunction of the donor organ in liver transplantation. Med J Australia 1991; 155:51-53.
96. Renard TH, Shimaoka S, Le Bherz D et al. ABO incompatible transplantation in children: A prospective approach. Transplant Proc 1993; 25:1953-1956.
97. Reding R, Veyckemans F, de Goyet JDV et al. ABO incompatible liver allografting in urgent indications. Surg Gyn Obst 1992; 174:59-64.
98. Gugenheim J, Samuel D, Reynes M et al. Liver transplantation across ABO blood group barriers, Lancet 1990; 336:519-523.
99. Gordon RD, Fung JJ, Markus B et al. Liver transplantation across ABO blood groups. Surgery 1986; 100:342-348.

CHAPTER 7

Liver Support System Development

Jacek Rozga, Kyung.S. Suh, Helene Lilja, Walid S. Arnaout, Achilles A. Demetriou

Historical Perspective

For many years it has been assumed that the majority of toxins which cause coma in hepatic failure are small dialyzable molecules. As a result, most liver support systems and therapeutic regimens relied primarily on blood detoxification. However, the pathogenesis of acute liver failure is complex and many investigators, including ourselves, have argued that isolated viable hepatocytes should be used to construct a liver support system to provide not only detoxification, but also missing liver synthetic functions. This suggests that a clear distinction should be made between systems designed to strictly "purify" patient blood from toxins and devices incorporating liver tissue preparations to carry out detoxification and provide synthetic function. The former are based on the artificial kidney principle; a classic example is hemodialysis, which was first applied by Kiley et al, and hemoperfusion, which was first tried by Schechter et al using cationic resins and several years later by Yatzidis and Chang et al using activated charcoal.[1-4] The latter include biological components (isolated hepatocytes, liver tissue slices) and are thus truly "bioartificial". In our opinion, the prefix "hybrid" should be used when both living, e.g., isolated hepatocytes and mechanical, e.g., adsorptive column, parts are used in combination to design a liver support system.

Charcoal hemoperfusion has been used to treat severe acute liver failure with mixed results. Although there is clear experimental evidence that the technique has some beneficial effects, a controlled prospective clinical trial has failed to demonstrate significant clinical benefits of the technique.[5] Other methods that relied primarily upon blood detoxification showed limited success as well. It would appear that, due to the complexity and vast number of metabolic and physiologic functions provided by the liver, construction of an extracorporeal liver support system requires utilization of viable isolated hepatocytes rather than specific cell components or enzymes. Sorrentino was the first investigator to use the term "artificial liver" while describing a series of in vitro experiments in which he demonstrated that fresh liver tissue homogenates can metabolize salicylic and barbituric acids and ketone bodies and produce urea from ammonia.[6] At approximately the same time, Sugie and Hori advanced the concept of cross-hemodialysis between man and dog.[7] A case report was soon published by Kimoto, and studies on hemodialysis against various canine liver tissue preparations were published by Nose et al.[8-10] The first clinical use of a device loaded with isolated liver cells was by Matsumura fourteen years later.[11] Rapid recovery from hepatic coma was accompanied by a significant decrease in blood ammonia levels. In the meantime, use of liver cells to construct a liver assist system was being studied extensively by Eiseman et al.[12,13] During the past forty years, artificial liver devices utilizing liver tissue preparations and

Support of the Acutely Failing Liver, 2nd ed., edited by Achilles A. Demetriou and Frederick D. Watanabe.

isolated liver cells were described by Mikami et al,[9] Wolf and Munkelt,[14] Eiseman et al,[12,13,15,16] Nose et al,[10] Kawarabata et al,[17] Hager et al,[18] Takahashi et al,[19] Wada and Ohshim,[20] Kimura et al,[21] Lie et al,[22] Kawamura et al,[23] Demetriou et al,[24] Kasai et al,[25,26] Saito et al,[27] Uchino et al,[28] Yanagi et al,[29] Margulis et al,[30] Matsumura et al,[11] Jauregui et al,[31,32] Arnaout et al,[33] Shnyra et al,[34] Takahashi et al,[35] Nyberg et al,[36-39] Sussman et al,[40-42] Fremond et al,[43] Li et al,[44] and by Rozga et al.[45-49] Several new systems are currently in various stages of laboratory development, including the novel hepatocyte bioreactors of Flendrig et al,[53] Wang et al,[54] Suda et al,[55] Gerlach et al,[56] Ekevall et al,[57] Qiang et al,[58] Mazariegos et al,[59] and Sajiki et al.[60] Most of these devices were tested in vitro and in animals with chemically and surgically induced hepatic insufficiency. In some studies, significant levels of detoxification and liver support were achieved. Patient data, however, are scarce and most of these devices were never used clinically.

The Challenge

In the setting of severe, irreversible liver necrosis resulting in acute liver failure, it is easy to understand that patients will not recover until the diseased liver is removed and replaced with a healthy organ. Thus, in this group of patients, the role of extracorporeal liver support and other therapeutic modalities is to "buy time" until a new liver becomes available. There will also be patients who suffer from severe, but potentially reversible, liver injury whose livers can recover and regenerate. In these patients, liver support therapy can potentially serve as the definitive treatment.

The challenge in the management of severe acute liver failure is to develop a therapeutic regimen for a disease in which we are faced with several major limitations:

1. The disease has varying etiologies and in reality we may be dealing with a group of diseases;
2. There is no clear understanding of the pathophysiology of the disease, thus making it difficult to develop rational and specific treatment strategies;
3. There is no single, accurate, "global" marker of quantifying liver functional mass and, as a consequence, the degree of liver dysfunction;
4. It is difficult to stratify the severity of illness and as a result determine the efficacy of therapeutic intervention;
5. Small numbers of patients with severe acute liver failure are seen at a single institution, making standardized data collection and development of management protocols difficult;
6. Progression of the disease is fairly rapid, with only a limited "window of opportunity" for initiation of therapy.

Past Accomplishments: Laboratory Studies

In Vitro Experiments

The results of in vitro experimental studies are summarized in Table 7.1. It appears that protein synthesis, gluconeogenesis, ureagenesis and both phase I, i.e., cytochrome P-450 system and phase II, i.e., conjugation, activities can be carried out by hepatocytes in these systems. This is encouraging, especially if one considers that cells are maintained in a non-physiologic environment away from numerous regulatory mechanisms which normally control their function in vivo.

In Vivo Animal Experiments

The in vivo experimental animal experience with hepatocyte-based artificial liver support systems is summarized in Table 7.2. It is difficult to assess and compare the efficacy of the

Table 7.1. Bioartificial liver in vitro

Author	Hepatocytes	Bioreactor	Effect
Wolf and Munkelt (1975)	Reuber hepatoma	hollow-fiber	bilirubin uptake and conjugation
Hager et al (1978)	mouse	hollow-fiber	ureagenesis, protein synthesis, diazepam metabolism, cytidine deamination
Kasai et al (1983)	dog	hollow-fiber	cytotoxicity of post-perfusion pig serum to dog hepatocytes, maintenance of ATP
Demetriou et al(1986)	rat, microcarrier -attached	chromatography column	bilirubin synthesis and conjugation, protein synthesis
Yanagi et al (1989)	rat and rabbit, gel-entrapped	rotating discs	ammonium removal, urea synthesis
Moscioni et al (1990)	human, micro-carrier attached	hollow-fiber	cyclosporine metabolism
Shatford et al (1992)	rat	hollow-fiber	albumin synthesis, amino acid and lidocaine clearance, oxygen consumption
Sussman et al (1992)	C3A (hepatoma -derived) cells	hollow-fiber	glucose utilization, albumin synthesis
Nyberg et al (1993)	rat	hollow-fiber	synthesis of albumin urea and ornithine, 0_2 uptake, lidocaine metabolism, arginine clearance
Li et al (1993)	rat	pyrex glass beads in a glass vessel	urea and albumin synthesis
Rozga et al (1993)	rat	hollow-fiber	cyclosporine and19-nortestosterone metabolism
Fremond et al (1993)	rat, gel (alginate) entrapped	Petri dish	bilirubin conjugation
Rozga et al (1994)	pig, aggregates	hollow-fiber	lidocaine, ethoxyresorufin metabolism
Jauregui et al (1994)	rat	hollow-fiber	diazepam metabolism
Bader et al (1995)	rat	bilaminar membrane bioreactor	cyclosporine and rapamycin metabolism
Gerlach et al (1996)	pig	woven multicompartment capillary system	amino acid and alpha-keto acid metabolism
Pazzi et al (1996)	pig	hollow-fiber	cholate metabolism

Table 7.2. Bioartificial liver in vivo

Author	Hepatocytes	Bioreactor	Animal Model	Effect
Matsumura et al (1973)	rat	hemodialyzer	CCl_4-intoxicated rats	clearance of BSP, bilirubin, NH_3, albumin synthesis
Olumide et al (1977)	pig	parallel plate membranes	anhepatic pig	neurologic improvement
Kasai et al (1985)	dog	hollow-fiber	galactosamine-intoxicated dogs	improved survival time
Uchino et al (1988)	pig	glass plates	anhepatic pig	improved survival time
Yanagi et al (1989)	rabbit, gel -entrapped	rotating discs	ischemic liver failure in cats	ammonium clearance
Arnaout et al (1990)	rat	Hollow-fiber	hyperbilirubin-emic (Gunn) rat	Bilirubin conjugation
Shnyra et al (1991)	rat, microcarrier-attached	unspecified column	CCl_4 and galactosamine-intoxicated rats	Improved survival
Williams et al (1991)	dog, microcarrier-attached	hollow-fiber	ischemic liver failure in dogs	NH_3 removal, improved glucose level, hemodynamic stability
Nyberg et al (1992 and 1993)	rat	hollow-fiber	anhepatic rabbit	albumin synthesis, clearance of amino acids, NH_3 and lactate
Sussman et al (1992)	C3A (hepa-toma-derived) cells	hollow-fiber	acetaminophen-intoxicated dogs	66% survived
Takahashi et al (1992)	porcine	50 glass plates	anhepatic rabbits	improved survival time ?
Fremond et al (1993)	rat	hollow-fiber	Gunn rat	bilirubin conjugation
Rozga et al (1993)	dog, pig, cryopreserved and fresh, microcarrier -attached	hollow-fiber	ischemic liver failure in dogs	normoglycemia, clearance of NH_3, lactate, normal PT, hemodynamic stability
Gerlach et al detoxyfication, (1994)	pig	woven multi-compartment capillary system	anhepatic pig	ammonia phenylalanine and lactate metabolism
Jauregui et al (1995)	rabbit	hollow-fiber	galactosamine-intoxicated rabbits	diazepam/lidocaine clearance, increased survival time

Table 7.2. Bioartificial liver in vivo (cont'd)

Author	Hepatocytes	Bioreactor	Animal Model	Effect
Dixit et al time (1996)	pig, encapsulated	conical chamber	galactosamine-induced liver failure	increased survival
Chen et al time (1996)	pig, cryopreserved aggregates	hollow-fiber	anhepatic pigs	increased survival
Suk et al survival (1997)	rat	hollow-fiber	fulminant hepatic failure; FHF	improved PT and time, accelerated HGF, c-Met, STAT3 and HNF1 expression in the FHF liver
Khalili et al time, (1998)	pig, cryopreserved aggregates	hollow-fiber	liver dearterialization in PCS-shunted pigs	increased survival prevention of brain edema

various systems described here. Most of the experimental animal models used in these studies did not reproduce the full range of metabolic and physiologic derangements seen clinically in severe acute liver failure. In addition, there was significant variation in the physiologic, biochemical and survival effects noted in the untreated groups. The baseline variation in outcome, seen in control animals, makes evaluation of the effects of the various interventions difficult. The major experimental animal models studied include those utilizing hepatotoxins, i.e., acetaminophen, CCl_4, D-galactosamine, and those with acute transient or permanent liver ischemia, sometimes in combination with partial liver resection. In both instances, however, it is difficult to standardize the degree of liver injury. Complete liver devascularization, although it appears to be more consistently reproducible, it is followed by a degree of rapid and severe deterioration of all vital functions and sepsis, which is rarely seen clinically. Thus, what can be achieved with artificial liver support using the devascularized liver animal model is at best a decrease in the rate of deterioration of various metabolic and physiologic parameters, i.e., acidosis, hypoglycemia, hyperammonemia and coagulopathy. Theoretically, the "purest" animal model for testing artificial liver support function is the anhepatic animal, because any liver-specific function detected during the study would be attributed solely to the support system. Therefore, this model would be useful in studies attempting to quantitate the "liver functional mass" of a specific liver support system. The anhepatic animal, however, does not reproduce the clinical setting of severe acute liver failure, i.e., massive liver necrosis and cerebral edema, and it cannot be used to predict how a specific liver support system will function in the acute liver failure environment.

Despite their limitations, experimental animal studies have shown that a number of metabolic functions, which are either impaired or lost as a result of severe liver insufficiency, can be replaced to some extent by liver support systems utilizing isolated hepatocytes and liver fragments (bioartificial livers). The first comprehensive studies utilizing a bioartificial liver were performed by Mikami et al and Nose et al.[9,10] At that time, enzymatic digestion of the liver had not been described; various canine liver tissue preparations (homogenates, fresh liver slices, freeze-dried granules of liver tissue) were being utilized. These investigators reported beneficial effects on glucose homeostasis, hyperlactemia and hyperammonemia. Soyer et al described a liver-assist extracorporeal perfusion system which utilized liver slices and, in an anhepatic pig model, demonstrated normalization of electroencephalographic tracings, presence of ureagenesis and formation of bilirubin conjugates.[15]

Use of isolated liver cell preparations to construct liver assist systems has been extensively studied by Eiseman et al,[12,13,16] who developed several liver cell-based systems, including a centrifuge, a dialyzer and a perfusion chamber. Eiseman's approach represents the most innovative work in the area of liver support system design and introduced two new important concepts:

1. Use of plasma separation, rather than whole blood perfusion, thus avoiding many of the complications associated with whole blood perfusion;
2. Placement of liver cells in a chamber within a high flow plasma recirculation loop. This recognizes the need for high flow to achieve effective exchange of nutrients, metabolites and oxygen between cell and plasma compartments.

In addition, Eiseman and his colleagues studied membrane hemodialysis in which suspensions of isolated liver cells circulated against blood removed from anhepatic pigs. Further pioneering work in this area was carried out by Uchino et al; they utilized high flow plasma recirculation and cultured liver cells on a matrix of collagen-coated borosilicated glass plates.[28] Unfortunately, neither Eiseman nor Uchino reported clinical testing of their systems. More recently, several groups of investigators attempted to support anhepatic rabbits and pigs using hepatocyte-based devices.[35,38,61,62]

Past Accomplishments: Clinical Studies

Clinical experience with liver support systems is limited. Takahashi et al reviewed the history of artificial liver development and described early clinical applications of devices utilizing hemodialysis against liver tissue homogenates, fresh liver slices and freeze-dried granules of liver tissue.[49]

As mentioned, the first clinical report of use of a liver support system utilizing isolated hepatocytes is that of Matsumura, who converted a dialyzer to an "artificial liver" by adding a cryopreserved rabbit liver cell suspension to the dialysis chamber.[11] In addition, he replaced the usual cuprophane membrane with a cellulose membrane permeable to middle range molecules, but not proteins. A 45-year-old man in hepatic failure due to unresectable cholangiocarcinoma underwent hemodialysis for 5 hours. Total serum bilirubin was reduced from 25.0 mg/dl to 16.8 mg/dl after the first treatment. Three days later, after a second 4 1/2 hour treatment, the total bilirubin level decreased from 18.0 mg/dl to 8.0 mg/dl. There was no evidence that the device affected the course of liver failure and outcome in this patient and there was no follow-up report on use of this system in a larger number of patients.

A group of 59 patients with liver failure was treated in Latvia with daily six hour hemoperfusions through a 20 ml capsule filled with porcine hepatocytes and activated charcoal particles.[30] Cells in the system were replaced at hourly intervals by fresh cells. Another group of 67 patients underwent standard medical therapy. The authors noted improved survival in the treated group; it was attributed to the support system. However, this study does not present convincing data to substantiate the severity of liver failure and it does not demonstrate evidence that the support system was actually functional. In addition, it appears that the number of cells used for each treatment was extremely small (less than 1 gram).

Sussman et al developed an extracorporeal liver assist device (ELAD) which utilized cells from a clonally-derived hepatoma cell line of human origin. Cells were grown rapidly to confluence and were well differentiated and strongly contact-inhibited. Early experiments performed in dogs with acetaminophen-induced liver failure, as well as ELAD treatments in a limited series of liver failure patients, were encouraging.[40-42] However, results of a pilot controlled study performed in two groups of patients with acute liver failure—those who had a potential for recovery and in those listed for emergent transplantation—were negative.[63]

We have developed a bioartificial liver (BAL) consisting of cryopreserved, matrix-anchored porcine hepatocytes and carried out a phase I clinical trial in patients with fulminant hepatic failure (FHF). The results of this study will be described in detail in subsequent sections of the monograph. However, it is important to note that the system was well tolerated and significant beneficial effects were observed in all treated patients.[64]

It can be easily seen that, despite several attempts to treat patients with severe liver failure using various liver support systems, there is only limited objective evidence that any of these systems had significant effects on the outcome of the underlying disease. Recent technological advances in cell isolation, culture and cryopreservation techniques, improved understanding of hepatocyte-matrix interactions, availability of new biomaterials and hollow-fiber bioreactors and better understanding of flow and mass transport across membranes have resulted in the development of a new generation of liver assist devices. Both in vitro (Table 7.1) and in vivo (Table 7.2), data utilizing these devices are currently being reported in the literature. There is clear evidence that hepatocytes remain functional and viable in these systems for relatively long periods of time and are able to provide metabolic and physiologic support in animal systems of liver failure.

Current Status of Liver Support

In this segment of the monograph, we will describe current work in the area of liver support, drawing mainly upon our own experience. We will describe experimental methodology developed by others and extensively used and/or modified by us and methods developed and presently used in our laboratory. Specifically, we will discuss:

1. Cell isolation, hepatocyte enrichment and cryopreservation techniques;
2. Liver support systems developed in our laboratory, defining their important engineering design and functional characteristics;
3. Experimental animal models used to test the function of liver support systems;
4. Critical elements emerging from the in vivo liver support system studies.

Hepatocyte Harvest and Isolation

Rat Hepatocytes

Donor rat hepatocytes were harvested using a modification of the collagenase digestion method developed by Berry and Friend.[50] Using general anesthesia, midline laparotomy was performed and the inferior vena cava was isolated and ligated. An 18-gauge catheter was placed into the portal vein and connected to an infusion pump. The liver was initially perfused with oxygenated 2 mM EDTA solution according to the method of Wang et al.[51] The perfusate was delivered at 25 ml/min and after 10 minutes it was changed to calcium-enriched 0.05% type IV collagenase. Collagenase perfusion was carried out at 37°C for another 10 minutes. The liver was then removed and the capsule incised. The digested liver tissue was suspended in Dulbecco's modified Eagle's medium (DMEM) and passed through a nylon mesh. Liver cells were washed three times in DMEM and centrifuged. Hepatocyte viability was determined by trypan blue exclusion. The final cell suspension contained > 80% viable hepatocytes and was further purified on a Percoll density gradient.[52]

Porcine Hepatocytes

Porcine hepatocytes were isolated from male and female pigs weighing 10-12 kg. Under ketamine anesthesia (30 mg/kg, i.v.), the abdomen was entered through a midline incision, the hepato-duodenal ligament was dissected and all its structures ligated and divided, except the portal vein which was cannulated with silicone tubing. The liver was perfused in situ with 3 l of a 2 mM EDTA solution prepared according to Wang et al at 300 ml/min using a roller pump.[51] Ten minutes later, the liver was resected and placed in a sterile basin which contained 0.1% collagenase type IV solution prepared in Ca^{2+}-enriched buffer. The solution was recirculated through the liver (300 ml/min) following passage through silicone tubing submerged in an oxygen-saturated water bath at 38°C. Twenty-five minutes later, the liver capsule was disrupted and digested liver parenchyma suspended in a large volume of ice-cold DMEM/10%BCS. Suspended liver cells were then passed through a nylon mesh and hepatocyte-enriched fractions isolated using a cell washer (COBE 2991); approximately 98% enrichment was achieved. Hepatocyte viability, assessed using trypan blue exclusion, was > 90 %; the total number of viable, purified hepatocytes was approximately 2×10^{10}.

Canine Hepatocytes

Male and female mongrel dogs (10-20 kg) were anesthetized with pentobarbital (30 mg/kg i.v.) and placed on a ventilator. Following total hepatectomy, the liver was placed in a sterile basin on ice cold saline and the portal vein was cannulated. The liver was perfused for 5 minutes with normal saline followed by perfusion, for 30 minutes, with a 2 mM EDTA solution prepared according to Wang et al.[51] The perfusate was passed through gas-permeable silicone tubing (Baxter Healthcare, McGaw Park, IL), submerged within a water bath maintained at 38°C and aerated with 100% oxygen. The oxygenated solution was perfused through the liver at 100 ml/min using a roller pump (MasterFlex, Cole-Palmer, Chicago, IL). The liver was then perfused with a solution similar to the above, minus EDTA and $NaHCO_3$, plus 1 mM $CaCl_2$ for 10 minutes. The pale, soft portions of liver were excised, minced and passed through a nylon mesh. Hepatocyte-enriched fractions were prepared by sedimentation (50 x g, 2 minutes). Viability was determined by trypan blue exclusion and consistently found to be greater than 60%.

Human Hepatocytes

In early studies, human hepatocytes were isolated by collagenase perfusion as follows. Portions of human liver, usually from the left lobe, weighing 300-400 g, were obtained from human organ donors who were not suitable liver donors. Catheters (#14 gauge) were introduced into all visible vessels on the cut surface of the liver fragment and sutured in place. Perfusion was carried out in a humidified chamber maintained at 37°C, with oxygenated perfusion buffer pumped through the liver fragment at a rate of 10-15 ml/min per catheter. The liver fragment was first perfused for 15 minutes with buffer without calcium. This was followed by perfusion with buffer containing 1.8 mM $CaCl_2$ and collagenase type I (Sigma, St. Louis, MO; 60 U/ml) for 20 minutes. Subsequently, the fragment was flushed with buffer containing 1.8 mM CaCl2 for 10 minutes. The pale digested regions of the liver were excised, minced and passed through a nylon mesh. Hepatocyte-enriched fractions were obtained by centrifugation (50 x g for 2 min) and washed three times with cold buffer. Cell viability was determined by trypan blue exclusion and found to be greater than 80%. Percoll gradient was used, as needed, to eliminate cellular debris and nonviable cells.[52]

Recently, we have modified our previous technique and used it to isolate hepatocytes from livers rejected for transplantation in our area. The intact liver was placed in a large 4 l metal bowl for surgical manipulation. The suprahepatic vena cava was oversewn and the cut end of the infrahepatic vena cava was surrounded with a 4-0 Prolene suture to close it snugly around a size 18 Masterflex tube (Barnant Co., Barrington, IL) inserted to provide the outflow from the liver. The hepatic artery and common bile duct were ligated using 0 silk ligatures. The portal vein was then identified and a size 16 Masterflex tube was inserted and secured with two 0 silk ligatures. This tubing acted as the hepatic inflow for the liver perfusion. The container with the liver was then placed in a water bath set at 42°C. The liver was permitted to float in 2 to 3 l of warm (37°C) physiologic saline. The tubing placed in the portal vein was attached to a roller pump with a loop situated in the water bath, while the end was placed in the reservoir of the perfusate. The roller pump speed was 0.2 ml/min/g liver tissue and the perfusate was placed below the level of the liver. The placement of the end of the outflow tubing from the liver depended upon the perfusate step. To begin the perfusion, 4 l of a 2 mM EDTA solution was flushed through the liver while the outflow was collected in a waste container placed below the level of the liver. Upon completion of the EDTA perfusion, 200 to 300 ml of calcium-enriched 0.05% collagenase type B (Boehringer Mannheim, Indianapolis, IN) solution was perfused through the liver and discarded. The outflow tubing from the infrahepatic inferior vena cava was then placed into the container with the collagenase solution, creating a closed recirculation circuit. Approximately 2 l of the collagenase solution was recirculated through the liver for 15-20 minutes until the liver was palpably digested. Once it was determined that the digestion was satisfactory, the tubing was removed from the liver and the saline was discarded from the bowl. Using sterile technique, 1 l of a 4°C solution containing Williams E (Gibco, Grand Island, NY) and Ham F12 (Gibco) at 1:1 (with 10% fetal bovine serum, 10 mol/l dexamethasone, 10.7 mol/l insulin) was slowly added while the liver capsule was manually disrupted, allowing the digested liver to fall freely into suspension. The cell suspension was then filtered via a custom-made filtration device with filters of decreasing porosity. The filtered hepatocytes were then washed with the Williams E/Ham F12 medium in a Cobe 2991 blood cell processor. The cell suspension was washed 3 times at 50 x g. The cells were resuspended in Williams E/Ham F12 and viability was determined by trypan blue exclusion; it was 68 ± 2% (n = 8). These cells were subsequently shown to metabolize significant amounts of diazepam and lidocaine.

Hepatocyte Attachment to a Matrix

Hepatocyte culture is facilitated by attachment to a collagen matrix. This allows maintenance of viable and functional hepatocytes for in vitro studies. We have used collagen-coated microcarriers as a convenient type of three-dimensional matrix for hepatocyte attachment. It allows culture, cryopreservation, shipping, transfer and placement of cells into bioreactors while attached to a solid surface. Many other types of matrix are available and essentially investigators will have to develop systems meeting their needs. In our laboratory, we attach 1.0-1.5 x 10^8 rat or canine hepatocytes and 1 x 10^9 porcine or human hepatocytes to hydrated collagen-coated dextran microcarriers by incubating them 3-24 hours at 37°C in 5% CO_2 in a variety of media in the presence of 10% fetal bovine serum.[24]

Hepatocyte Cryopreservation

After incubation, microcarrier-attached hepatocytes were stored at -70°C in DMEM/10% fetal bovine serum and 10% (v/v) dimethylsulfoxide for periods of up to 6 months. After cryopreservation, the viability of microcarrier-attached porcine hepatocytes was at least 65%. However, functional assessment of thawed cryopreserved hepatocytes from various species gave mixed results. Further work in this important area is needed for optimizing hepatocyte cryopreservation techniques to allow maintenance of liver-specific differentiated functions.

In Vitro Liver Support System Studies

Cytochrome P-450 Functions

We carried out in vitro experiments to determine whether microcarrier-attached hepatocytes inoculated in a hollow-fiber module can carry out cytochrome P-450-specific functions. The ability to carry out such functions is essential for successful use of an extracorporeal liver support system. The experimental set-up consisted of 30 ml modules (Minikap/225, Microgon, Inc.). Each module was inoculated into the extrafiber space with 3 x 10^7 freshly thawed microcarrier-attached rat hepatocytes. One hour later, either cyclosporine A (CyA) or 19-nortestosterone (19-NT) was inoculated into the system (1 mg/ml). In control experiments, either vehicle alone or a test drug was injected into a system containing microcarriers alone. CyA and its metabolites were extracted and analyzed by high performance liquid chromatography (hplc) using a Hewlett-Packard Model 1090 chromatograph with workstation (Avondale, PA). CyA metabolites were rechromatographed on a LC-C8 analytical column. 19-NT and its metabolites were extracted and submitted to gas chromatography/mass spectrometry (GC/MS) analysis. For the determination of the glucuronides of 19-NT, samples were dried and hydrolyzed. Analyses were performed using a Hewlett-Packard model 5988A Gas Chromatograph/ Mass Spectrometer (Avondale, PA) fitted with a Hewlett-Packard HP1 methyl-silicon column (16.6 mm x 0.2 mm; 0.33 m film thickness).

CyA metabolites were isolated from all modules containing microcarrier-attached hepatocytes following addition of CyA. In control modules with microcarrier-attached hepatocytes not receiving CyA, hplc peaks corresponding to its metabolites were not detected; similarly, CyA metabolites were not detected in modules containing microcarriers alone. Analysis of module extracts demonstrated two prominent peaks (X and Y) with UV spectral characteristics of CyA and retention times corresponding to known metabolite standards. Both peaks increased linearly with time after four hours of incubation. Peak "X" increased approximately 2-fold (rate: 2.8 mg/1 x 10^7 cells/hour) and accounted for 8.5% of the original dose at eight hours of incubation; peak "Y" increased approximately 1.5-fold (rate: 1.0 mg/ 1x10^7 cells/hour) and accounted for 3.1% of the original CyA dose at the end of eight hours. When peaks X and Y were re-chromatographed, peak X resolved into three metabolites which eluted at retention times of 14, 15 and 16 minutes respectively, and peak Y resolved into three metabolites which eluted at retention times of 21, 22 and 23 minutes respectively. Comparison of this hplc pattern to the pattern obtained from extracted serum from a patient undergoing CyA therapy demonstrated production of identical metabolites. Based on HPLC retention and UV-spectral assignments times of these metabolites, we concluded that the three metabolites isolated from peak X represented first generation mono-hydroxylation products of CyA, two of which would be metabolites 17 and 1, and that the metabolites under peak Y were first generation N-demethylation products, one of which would be metabolite 21. Metabolites of 19-NT were identified in the media after inoculation and increased in concentration with time. The predominant metabolite was the glucuronide of 19-NT, which increased at an approximate rate of 134 ng/1 x10^7 cells/hour and accumulated, after 8 hours of culture, to about 3.5% of the level of the parent compound administered. The remaining metabolites were isomerization products of the parent compound and amounted to less than 1% of the administered 19-NT at the end of the experiment. Identified from media following six hours of culture were: 19-NT, 19-noretiocholanolone and 19-norepiandrosterone.

Hepatocyte Calorimetry

To further characterize microcarrier-attached hepatocytes metabolically, experiments were carried out to establish a tissue culture indirect calorimetry method. The calorimeter used (Micro-Oxymax, Columbus Instruments, Columbus, OH) consisted of an automated closed system in which air is pumped through gas sensors and then returned to a cell culture flask. The sensitivity of the system is 0.01-0.03 ml/min. Fresh adult rat hepatocytes (5×10^7) isolated by collagenase were suspended in 20 ml of DMEM and placed in a sterile 50 ml water-jacketed glass flask and stirred gently. The temperature was maintained at 37°C by circulating water. Oxygen consumption and CO_2 production were recorded every 90 minutes for seven hours. The respiratory quotient (RQ) was calculated. At three hours, we found the following: O_2 consumption: 3.45 ± 0.59; CO_2 production: 10.75 ± 1.29; RQ: 3.60 ± 0.55. At seven hours, there was a decrease in cell viability by 30% coinciding with a decrease in O_2 consumption (1.74 ± 0.72) and CO_2 production (4.04 ± 0.36) and an increase in RQ (6.0 ± 2.1) suggesting a shift towards anaerobic metabolism. Cryopreserved hepatocytes had lower baseline O_2 consumption and CO_2 production. It thus appears that indirect cell calorimetry can be used to determine the metabolic state of cultured hepatocytes and monitor their viability in vitro. Experiments are in progress to examine its usefulness in monitoring cell viability and metabolic function in liver support systems.

Hepatocyte Proliferation

The ability of normal adult hepatocytes to proliferate in response to a known mitogen, epidermal growth factor (EGF), was studied. Adult male rat hepatocytes were obtained by in situ portal vein collagenase perfusion, attached to collagen-coated microcarriers by incubation in DMEM/10%FBS (5% CO_2/95% air, 37°C) for 2.5-24 hours and then inoculated into hollow-fiber bioreactors at a density of 4×10^7 viable cells/bioreactor. Bioreactor cultures were perfused with media (DMEM/5% FBS; 150 ml/bioreactor) at 2 ml/minute under 5% CO_2 and 95% air at 37°C. Three hours after culture initiation, EGF was added to media at levels of 0, 10, and 50 ng/ml. After two hours exposure to EGF, ^{3}H-thymidine was added to all media reservoirs (2 mg/ml) and culture continued for an additional three hours. Cells were then removed from the bioreactors and extracted for total DNA. Measurements were made for total cellular DNA and ^{3}H-thymidine incorporation into DNA. ^{3}H-thymidine incorporation into DNA of cultured cells (dpm/mg DNA) increased significantly (control: 2.28 ± 0.12; 10 ng/ml EGF: 3.45 ± 0.22; 50 ng/ml EGF: 3.58 ± 0.40) with both doses of EGF ($p < 0.02$). No significant differences in total cellular DNA and protein contents were noted.

In Vivo Animal Liver Support Studies

Treatment of a Genetic Defect in a Small Animal Model

Cryopreserved microcarrier-attached rat hepatocytes ($3\text{-}4 \times 10^7$) were inoculated into the extracapillary chamber of a hollow-fiber bioreactor. The BAL containing either microcarrier-hepatocytes (n = 6) or microcarriers alone (n = 5), was attached to adult hyperbilirubinemic Gunn rats via cannulas in the aorta and inferior vena cava. Bile samples were collected before and at hourly intervals after attachment to the BAL, after cannulation of the bile duct. Neither plasma separation nor pumps supporting blood flow was used. The same animal model was used to demonstrate de novo conjugation of free bilirubin by the BAL (n = 3) after systemic administration of unconjugated ^{3}H-bilirubin. Bile was analyzed for bilirubin mono- and di-glucuronides by high performance liquid chromatography (hplc) and results were confirmed by digestion with β-glucuronidase and co-chromatography with normal rat bile. A progressive increase in the concentration of bilirubin mono-glucuronide and bilirubin di-glucuronide in bile was seen as early as thirty minutes and lasted up to four hours following animal attachment to a BAL containing microcarrier-normal hepatocytes (total bilirubin glucuronide concentration: 3.53 ± 0.68 to 8.07 ± 0.85 mM; $p < 0.01$). Approximately 22% of the radiolabeled bilirubin excreted in bile was seen in the form of bilirubin conjugates; no radioactivity was seen in the free bilirubin fraction. We conclude that these data demonstrate function of the BAL in this animal model.

Development of Large Animal Model of Liver Failure

As already mentioned, one of the major problems in carrying out studies to assess the efficacy of extracorporeal assist devices is lack of reproducible animal models of acute liver failure. Another limitation is the difficulty in carrying out "survival" studies in large animals. We have developed a large animal experimental model for acute liver failure; it is reliable, reproducible, well defined and humane. Initially, under intravenous pentobarbital,[10] adult mongrel dogs underwent end to side portocaval shunt and placement of exteriorized ligatures around the common hepatic and gastroduodenal arteries. The crucial difference of this technique from other previously described ones is the ligation of the gastroduodenal artery. We found that retrograde flow through this vessel, following ligation of the hepatic artery proper in these animals, was associated with recovery of liver function. This could account for inconsistent results obtained with that model in the past. Animals were allowed to recover for 48 hours, at which time, they were re-anesthetized and supported by mechanical ventilation. Jugular venous and femoral arterial cannulas were placed for fluid maintenance and pressure monitoring respectively. The exteriorized arterial ligatures were applied after collection of baseline blood for serum glucose, ammonia, lactate, transaminases and pH determinations. Each animal included in the study was administered 0.5 mg/kg indocyanine green (ICG) and its clearance was determined before shunting and at two hours following arterial ligation. Prior to the portocaval shunt, there was rapid clearance of ICG at twenty minutes (absorbance at 280 nm: 1.90 ± 0.08-0.52 ± 0.08); following arterial ligation, clearance was significantly decreased (absorbance at 810 nm: 2.0 ± 0.10-1.80 ± 0.09). At five hours following hepatic devascularization, there were significant metabolic derangements in these animals. Serum glucose decreased (94.7 ± 12.7 to 22.5 ± 4.31 mg/dl; $p < 0.02$), ammonia increased (6.43 ± 0.6 to 16.3 ± 0.8 mg/l; $p < 0.05$), lactate increased (1.91 ± 0.3 to 11.3 ± 1.4; $p < 0.05$) and pH decreased (7.34 ± 0.07 to 7.02 ± 0.05; $p < 0.05$) significantly. In addition, there was a significant increase in the activity of serum transaminases following hepatic devascularization. Random liver biopsies at the end of the experiments (6-7 hours) revealed massive liver necrosis.

Testing of BAL in Large Animals with Liver Failure

Animals were divided into three groups: Group I consisted of control dogs (n = 6) attached to a BAL inoculated with microcarriers alone. Group II dogs (n = 4) were attached to a BAL inoculated with cryopreserved microcarrier-attached allogeneic (canine) hepatocytes. Group III consisted of dogs (n = 6) attached to a BAL inoculated with cryopreserved microcarrier-attached xenogeneic (porcine) hepatocytes. Following cannulation of the femoral artery and vein, heparin (100 U/kg i.v.) was given to all animals and plasma separation was initiated with a Haemonetics 30-S blood processor (Haemonetics Inc., Braintree, MA). Following plasma separation, plasma was pumped through two parallel hollow-fiber modules at 30 ml/min. Approximately 80% of the flow was directed through the hollow-fiber lumen, while 20% crossed the porous cellulose acetate membrane to the extrafiber space and exited through a side port. Plasma was reconstituted with the blood cells before re-infusion into the animal (Fig. 7.1). Perfusions were carried out for six hours. Blood samples were obtained at 0, 60, 120 and 360 minutes; prothrombin time, glucose, ammonia, lactate and activity of LDH and AST determinations were carried out at the hospital clinical laboratory.

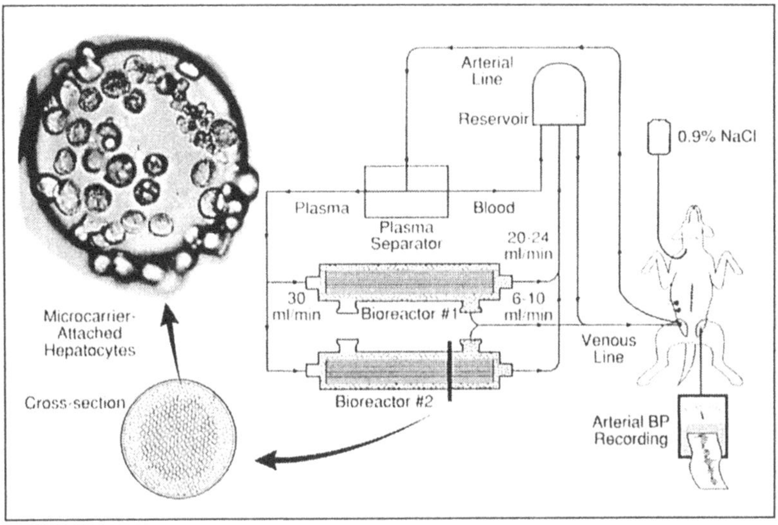

Fig. 7.1. BAL support system. Hollow-fiber modules were inoculated with microcarrier-attached liver cells.

Table 7.3. Systolic blood pressure in study animals

Group	Baseline	1 hr	2 hr	4 hr	6 hr
Controls	187 ± 12[a]	125 ± 10	106 ± 5	75 ± 10	50 ± 10
BAL (dog hepatocytes)	212 ± 12	162 ± 7[b]	131 ± 12	118 ± 6[b]	100 ± 10[b]
BAL (Pig hepatocytes)	188 ± 14	155 ± 14	142 ± 12[b]	124 ± 8[b]	187 ± 12[b]

[a] Data expressed as mean ± SD
[b] p=0.05 or less compared with controls

At the completion of the perfusion period, animals were killed with pentobarbital overdose. Attachment of the animals to a BAL system inoculated with cryopreserved microcarrier-attached allogeneic canine hepatocytes and xenogeneic porcine hepatocytes produced several significant changes. As shown in Table 7.4, after four hours of treatment, blood ammonia and serum lactate levels were significantly lower and serum glucose was significantly higher in both groups compared to control animals. In addition, the experimental groups of animals had significantly higher mean systolic blood pressure when compared to controls (Table 7.3). Differences in prothrombin time were not significant. As expected, serum LDH and AST levels progressively increased, similarly in all three groups (Table 7.4). Hepatocyte viability at the end of the six hour experimental period was greater than 90% in all groups.

Table 7.4. Laboratory values in in vivo BAL studies

Parameters	Baseline	1 hr	2 hr	4 hr	6 hr
Glucose (mg/dl)					
Control	102 ± 7[a]	87 ± 10	39 ± 9	9 ± 2	4 ± 0.2
BAL (dog hepatocytes)	142 ± 34	162 ± 31	120 ± 30[b]	76 ± 15[b]	48 ± 11[b]
BAL (pig hepatocytes)	117 ± 5	128 ± 17	89 ± 17	41 ± 11[b]	25 ± 7[b]
Ammonia (mg/dl)					
Control	95 ± 4	132 ± 8	149 ± 8	188 ± 11	220 ± 12
BAL (dog hepatocytes)	85 ± 11	132 ± 11	130 ± 8	140 ± 9[b]	186 ± 12
BAL (pig hepatocytes)	111 ± 11	146 ± 10	141 ± 9	137 ± 7[b]	140 ± 9[b]
Lactate (mmol/l)					
Control	21 ± 3	68 ± 16	83 ± 18	120 ± 25	116 ± 26
BAL (dog hepatocytes)	17 ± 3	52 ± 18	36 ± 11[b]	43 ± 6[b]	68 ± 10[b]
BAL (pig hepatocytes)	11 ± 6	22 ± 10[b]	27 ± 10[b]	27 ± 11[b]	33 ± 12[b]
LDH (mIU/ml)					
Control	57 ± 9	136 ± 26	288 ± 89	592 ± 305	1738 ± 481
BAL (dog hepatocytes)	182 ± 12	280 ± 55[b]	297 ± 43	403 ± 82	403 ± 181
BAL (pig hepatocytes)	126 ± 26	241 ± 39	325 ± 64	514 ± 95	741 ± 80
AST (mU/ml)					
Control	51 ± 12	48 ± 9	76 ± 21	333 ± 131	1174 ± 481
BAL (dog hepatocytes)	56 ± 8	135 ± 46	131 ± 46	223 ± 54	403 ± 181
BAL (pig hepatocytes)	62 ± 21	79 ± 44	100 ± 23	336 ± 91	188 ± 14

[a] Data expressed as mean ± SEM
[b] p=0.05 or less compared with controls

Development and Testing of a Hybrid BAL System

Of all the various extracorporeal support methods of treating severe liver failure, charcoal hemoperfusion has been most extensively studied both in animals and humans. The purpose of this study was to determine the efficacy of a novel hybrid system combining charcoal plasma perfusion with perfusion through a hollow-fiber module inoculated with normal hepatocytes in treating animals with irreversible severe liver failure, and compare it to that of charcoal plasma perfusion alone.

Microcarrier-attached porcine hepatocytes were inoculated into the extrafiber compartment of hollow-fiber modules (Fig. 7.1). Each hollow-fiber module (Z22M-060-01X; Microgon, Inc., Laguna Hills, CA) consisted of a polycarbonate cylinder (29.1 mm I.D., 31.2 mm O.D.) containing 670 cellulose nitrate/cellulose acetate fibers with an extrafiber volume of 177 ml. The total internal fiber surface area was 5,850 cm^2; external surface area was 7,010 cm^2 and pore diameter in the semi-permeable fiber wall was 0.2 m.

A column containing 150 g of cellulose-coated charcoal was used (Adsorba 150C, Gambro Dialysatoren GmbH & Co., KG, Germany). Prior to use, each column was saturated with glucose and washed with 2 l of heparinized saline. The large animal experimental model of ischemic liver failure described above was utilized. Following cannulation (by groin cut down) of the femoral artery and vein, animals were heparinized (100 U/kg, i.v.) and blood was removed through the venous cannula. Plasma was separated using a filtration column (Plasmaflo AP-05H; Asahi Medical Co., Apheresis Technologies, Inc., Palm Harbor, FL); at a blood flow of 150 ml/min, approximately 25-30 ml/min of plasma could be removed continuously without inducing thrombocytopenia and hemolysis.

The Liver Support System (Fig. 7.2) consisted of a plasma separator, a transmission reservoir, a charcoal column, a hollow-fiber module inoculated with 5 x 10^9 microcarrier-attached porcine hepatocytes, an oxygenated water bath maintained at 37°C, four roller pumps (Masterflex, Cole-Palmer, Chicago, IL) and oxygen-permeable silicone tubing (Baxter Healthcare, McGaw Park, IL). The primary mechanism of solute transport in a hollow-fiber module is fluid convection (Starling flow) driven across the membrane by a trans-fiber pressure drop. Therefore, the faster the axial flow rate, the faster the Starling flow. Since plasma could not be removed from the filtration column at a rate faster than 30 ml/min, a transmission reservoir was introduced into the system from which plasma recirculated in the BAL circuit at 220 ml/min. Following recirculation through the BAL, plasma and blood cells were reconstituted and blood returned to the animal at 30 ml/min via the venous cannula.

Two groups of animals were studied. Group I animals (n = 7) were treated with a BAL system consisting of a charcoal plasma perfusion column and a hollow-fiber module inoculated with 5-6 x 10^9 viable microcarrier-attached porcine hepatocytes. Group II animals (n = 6) were treated with charcoal plasma perfusion alone. The number of cells used in the BAL represented at least 50-60 g of the liver, i.e., at least 10% of the total estimated liver weight of the dog (dog weight: 23-30 kg). Perfusion of each anesthetized animal was carried out for six hours. Blood was collected before and at hourly intervals following hepatic devascularization for determination of glucose, electrolytes, ammonia, lactate, LDH, AST, haptoglobin,

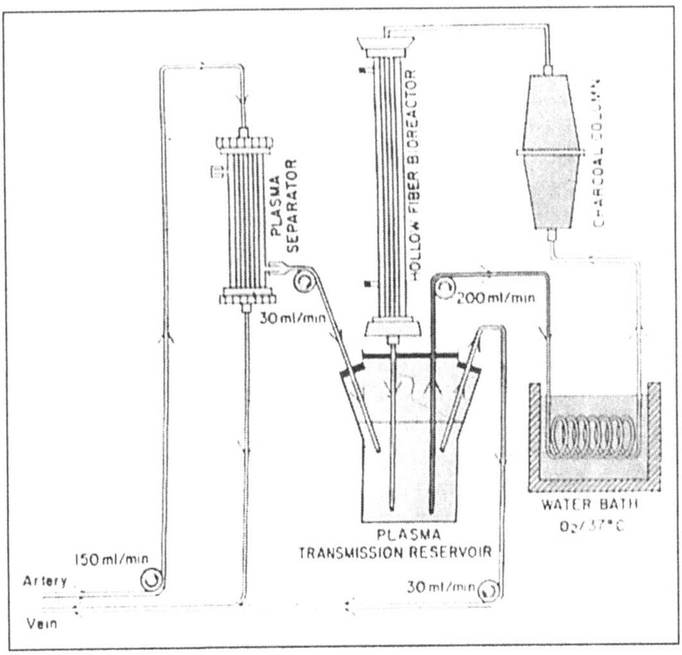

Fig. 7.2. Schematic representation of the bioartificial (BAL) support system.

prothrombin time, fibrinogen and platelet count. All assays were carried out at the hospital clinical laboratory utilizing standard methods. Two hours following hepatic devascularization, 0.5 mg/kg of indocyanin green (ICG) was administered intravenously and blood samples were obtained at five minute intervals for twenty minutes to determine ICG clearance. Plasma ICG levels were determined spectrophotometrically (absorbency at 810 nm). At the completion of the perfusion period, animals were killed with intravenous pentobarbital overdose.

Hepatocyte viability following harvesting with EDTA/collagenase portal vein perfusion was consistently greater than 90%. It remained unchanged following the incubation period and most cells (> 75%) were attached to the microcarriers. Cells were used immediately following the attachment period (24 hours).

No technical problems were encountered during either BAL or charcoal treatment. There was no hemolysis in both groups of animals; at the end of the six hour perfusion period, haptoglobin levels remained normal (<150 mg/dl). However, thrombocytopenia (platelet count less than 100,000 but greater than 50,000/ml) was seen in three out of the thirteen animals (one in Group I and two in Group II). In both groups of animals, as expected, there was a progressive, significant increase in the activity of serum LDH (Group I: 585 ± 24; Group II: 492 ± 109 mIU/ml) and AST (Group I: 348 ± 74; Group II: 317 ± 106 mU/ml) after six hours of treatment. The differences in enzyme values between the two groups were not statistically significant, suggesting similar degree of liver injury. At four, five and six hours of treatment, serum ammonia and lactate levels were significantly lower and serum glucose was significantly higher in Group I compared to Group II (Table 7.5). It is important to note that in Group I animals, following an initial rise in ammonia levels, there was no further increase and that four out of seven animals remained normoglycemic throughout the six hour treatment period. In contrast, Group II dogs became hypoglycemic early and had a progressive rise in serum ammonia levels (Table 7.5). At six hours, Group I dogs had a normal prothrombin time, whereas Group II dogs had a significantly prolonged prothrombin time. The difference in plasma fibrinogen levels between the two groups was not statistically significant. ICG clearance was over 20% at fifteen minutes and over 40% at twenty minutes in Group I dogs, whereas Group II animals showed near total dye retention (Fig. 7.3). Group I animals remained hemodynamically stable throughout the treatment period, with a mean systolic blood pressure of 105 ± 11 mmHg at six hours compared to Group II dogs (52 ± 8 mmHg; $p < 0.020$). Hepatocyte viability in the liver support module at the end of the six hour treatment was greater than 90%.

In summary, we have demonstrated that attachment of a hollow-fiber module inoculated with normal hepatocytes to Gunn rats (which are unable to conjugate bilirubin), via arterial and venous cannulas, resulted in the appearance of bilirubin conjugates in their bile. In subsequent experiments, utilizing dogs with severe irreversible acute ischemic liver failure, we demonstrated that treatment with a hollow-fiber module inoculated with cryopreserved matrix-attached allogeneic and xenogeneic (porcine) hepatocytes resulted in significant beneficial effects. In another experiment, the efficacy of a hybrid (charcoal column and hepatocyte module) liver support system was compared to that of a system utilizing charcoal alone. We hypothesized that by placing the charcoal column before the hepatocyte module in the circuit, we would not only enhance the BAL's detoxifying capability, but also possibly protect the normal porcine hepatocytes from the toxic effects of hepatic failure plasma.

To avoid hemolysis and platelet depletion associated with whole blood perfusion, a plasma separation system was used which allowed perfusion of the hepatocyte bioreactor with plasma only. In addition, it eliminated direct cell-cell interactions between host blood cells and xenogeneic liver cells. Finally, high-flow recirculation of plasma was used to improve bidirectional solute transport and oxygen delivery. Our results demonstrated that

Table 7.5. Laboratory values in dogs with acute liver failure treated with bioartificial liver (BAL) and charcoal plasma perfusion alone (charcoal)

Group	Baseline	1 hr	2 hr	3 hr	4 hr	5 hr	6 hr
Glucose	**(mg/dl)**						
BAL	99 ± 7	109 ± 7	97 ± 10*	75 ± 10*	69 ± 13*	60 ± 13*	50 ± 12*
Charcoal	100 ± 6	88 ± 10	53 ± 12	27 ± 14	17 ± 10	9 ± 4	4 ± 1
Ammonia	**(mmol/l)**						
BAL	95 ± 8	124 ± 11	170 ± 17*	168 ± 21*	170 ± 20*	172 ± 17*	164 ± 16*
Charcoal	105 ± 9	297 ± 89	286 ± 43	293 ± 57	377 ± 102	461 ± 140	440 ± 110
Lactate	**(mmol/l)**						
BAL	1.8 ± 0.2	3.4 ± 0.5	4.9 ± 1.0	5.1 ± 1.1	5.2 ± 1.2*	5.3 ± 1.2	5.4 ± 1.0
Charcoal	1.5 ± 0.1	4.3 ± 0.5	5.9 ± 0.5	7.7 ± 0.5	9.7 ± 0.5	11.6 ± 0.4	10.7 ± 0.9
Prothrombin	**Time (sec)**						
BAL	11.3 ± 0.5	10.6 ± 0.3	10.4 ± 0.2	10.7 ± 0.3	11.1 ± 0.6	11.6 ± 0.7	12.0 ± 0.6*
Charcoal	13.3 ± 1.1	11.0 ± 0.4	10.5 ± 0.3	11.2 ± 0.4	11.6 ± 0.6	13.4 ± 1.3	17.0 ± 2.3
Fibrinogen	**(mg/dl)**						
BAL	194 ± 18	159 ± 26	140 ± 26	126 ± 22	130 ± 26	110 ± 15	91 ± 16
Charcoal	155 ± 14	115 ± 20	99 ± 17	81 ± 19	77 ± 17	67 ± 14	63 ± 14

BAL: Group I
Charcoal: Group II
*p=0.5 or less, unpaired Student t-test

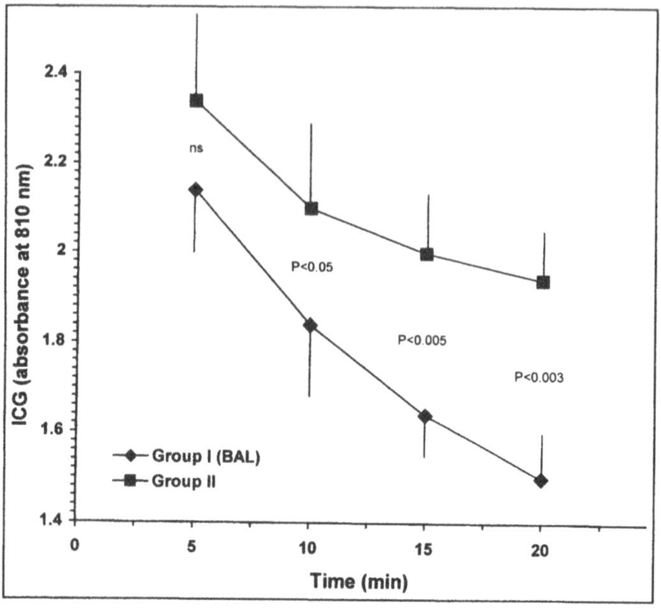

Fig. 7.3. Indocyanin green values (absorbence at 810nm) at 5, 10, 15 and 20 minutes after intravenous dye administration (0.5 mg/kg BW).

the BAL provided a significant level of detoxifying, metabolic and hemodynamic support. After six hours on the BAL system, hepatocyte viability and attachment to microcarriers remained high, suggesting that the extrafiber compartment of the hollow-fiber bioreactor was effectively perfused with oxygenated plasma. Animals tolerated BAL treatment well, and no hypersensitivity or other acute immunologic reaction was noted. This could be due to the immunosuppressive effect of severe liver failure. However, it is possible that after repeated BAL plasma perfusions, xenogeneic hepatocytes may either have a detrimental effect on the treated animals or could be destroyed by circulating antibodies in the plasma. If that is the case, techniques would have to be used (i.e., binding of circulating antibodies, immunosuppression, use of hollow fibers with molecular weight cut-off less than 100,000) to neutralize the antibodies.

Testing of a Hybrid BAL System in Anhepatic Pigs

We have reported reduction of brain edema and neurologic improvement in patients with fulminant hepatitis by either simultaneous or sequential use of a bioartificial liver (BAL) and induction of anhepatic state. However, wider use of this approach can be undertaken if the BAL is shown to provide effective metabolic and physiologic support during the anhepatic state. We examined this possibility by first developing an anhepatic pig model and then by studying the effects of temporary support with a BAL in pigs rendered anhepatic.

Adult farm pigs (45-60 kg) of either sex were used. Under general (isoflurane) anesthesia, the abdomen was entered through a long midline incision. An open cystostomy was performed to place a catheter for urine output monitoring. The portal vein and the infrahepatic inferior vena cava (IVC) were identified and dissected. An end to side portocaval shunt (PCS) was performed with 5-0 Prolene (Davis & Geck, Manati, PR) continuous suture. During the PCS anastomosis, care was taken to only partially occlude the IVC. The hepato-duodenal ligament was dissected to identify the proper hepatic artery and the common bile duct, which were ligated and divided. After applying vascular clamps to the infrahepatic and suprahepatic IVC, the liver was resected. The vena caval flow was reconstituted with an 18 mm polytetrafluoroethylene (PTFE) graft (Gore-Tex, Flagstaff, AZ) using 4-0 Prolene (Davis & Geck, Manati, PR) continuous suture. Using this technique, the two episodes of splanchnic venous stasis (PCS and IVC grafting) lasted only 10-15 minutes each and were separated by a 10 minute interval during ligation of the hepatoduodenal ligament. Heparin and veno-venous bypass were not utilized. The abdominal fascia and skin were closed separately. The animals were recovered from anesthesia and disconnected from mechanical ventilation. During the immediate postoperative period prior to the BAL treatment, glucose was administered intravenously by continuous infusion (2.0-3.0 ml/min). The IV fluid was changed to lactated Ringer's solution during the BAL treatment. Arterial blood gas was evaluated using a blood gas analyzer (ABL5, Radiometer Medical A/S, Denmark).

The BAL consisted of microcarrier-attached cryopreserved porcine hepatocytes (5×10^9) in a hollow-fiber module, a charcoal column and an oxygenator (HepatAssist 2000, CIRCE Biomedical Inc., Lexington, MA). Through a double-lumen catheter placed in the jugular vein, blood was removed and separated into plasma and blood cells (COBE Spectra). Plasma was directed into the BAL, where it was recirculated at 400 ml/min. Treatment with the BAL was initiated after the animal recovered from anesthesia and was carried out from the fourth to the tenth hour post-operatively. Animals were maintained normothermic through external warming (lamp, heating pad). Blood glucose levels were determined every 30 minutes and kept above 60 mg/dl with boluses of intravenous dextrose. Group 1 pigs (n = 8) were treated with the complete BAL and Group 2 pigs (n = 6) with the BAL containing charcoal alone. After the BAL treatment, pigs were maintained on intravenous lactated Ringer's solution with 5% dextrose (2.0-3.0 ml/min) until death.

At the start of treatment, all pigs had normal body temperature, blood pressure, heart rate, urine output, platelet count, serum electrolytes, total protein, glucose, creatinine and liver function tests. Abnormal readings included (Group 1 vs. Group 2) hyperammonemia (285 ± 145 vs. 339 ± 112 mM/l), lactic acidemia (4.6 ± 2.5 vs. 7.1 ± 4.9 mM/l) elevated alkaline phosphatase (394 ± 317 vs. 401 ± 129 U/l) and prolonged prothrombin (PT) time (17.1 ± 1.7 vs. 15.9 ± 1.3 sec).

Beneficial effects of treatment with the BAL (Group 1) as compared to treatment with charcoal alone (Group 2), included prolongation of survival, lower requirement for intravenous dextrose, lower reduction of fibrinogen, maintenance of PT (18.5 ± 0.8 sec.) and higher urine output (Table 7.6).

In summary, during the initial post-operative period pigs rendered anhepatic showed remarkable stability and maintenance of all vital functions. Temporary support with the BAL significantly prolonged survival time in these animals.

Testing of BAL in Rats with Fulminant Hepatic Failure (FHF)

We earlier described a model of FHF in the rat where partial hepatectomy is combined with induction of right liver lobe necrosis.[65] After this procedure, lack of regenerative response in the residual viable liver tissue was associated with elevated plasma hepatocyte growth factor (HGF) and transforming growth factor β (TGF-b1) levels and delayed expression of HGF and HGF receptor c-Met mRNA in the remnant liver. We next demonstrated that in FHF rats, intrasplenic transplantation of a small number of isolated hepatocytes (2% liver mass) prolonged survival, improved liver function tests and facilitated regenerative response in the remnant liver.[66] The latter effect was associated with accelerated expression of HGF and c-Met mRNA in the liver and lowering of plasma HGF and TGF-β1 levels. This study, which is described in greater detail in a section below, suggests that hepatocyte-based therapy may facilitate regeneration and repair of the native liver,

Table 7.6. Laboratory values in anhepatic pigs treated with a hybrid bioartificial liver (HepatAssist 2000TM)

Group	Survival (hr)	Dextrose intake (mg/kg/6 hrs)	Fibrinogen (reduction %)	Urine output (ml/kg/6 hrs)
BAL	22.0 ± 7.9	190 ± 173	-44 ± 14	60 ± 24
Charcoal	13.2 ± 1.6*	542 ± 186*	-67 ± 9*	37 ± 6*

BAL: Group 1
Charcoal: Group 2
Data are shown as means ± SD
* All differences at p=0.05 or less (paired Student's t-test).

presumably through lowering of a potent hepatocyte growth inhibitor such as TGF-β1. We therefore examined whether treatment of FHF rats with a bioartificial liver (BAL) utilizing isolated functional hepatocytes would produce similar effects.

Inbred male Lewis rats (350 g) were used. FHF was induced as described above. The BAL consisted of a hollow-fiber module (Microgon; 0.2 m pores, 300 fibers, 7.2 ml extrafiber space) loaded with 3 x 10^8 syngeneic microcarrier-attached (Cytodex 3; Pharmacia) hepatocytes. The sham-BAL contained microcarriers only. Group 1 FHF rats (n = 12) were treated with the BAL. In Group 2 FHF controls (n = 12), sham-BAL was used. At 4 hours after induction of FHF, the BAL circuit was filled with heparinized blood, submerged in a warm (38°C) water bath and attached to the femoral vessels. A pump-assisted whole blood perfusion was then carried out for 4 hours at 2 ml/min. During treatment, rats were placed in an oxygen-saturated cage so that the arterial PaO_2 was maintained above 200 mmHg. In 6 rats from each group, survival was determined; the remaining 12 rats were euthanized at 4 hours after BAL/sham-BAL treatment. At sacrifice, blood was collected for measurement of HGF and TGF-β1 levels (ELISA). The native liver was harvested and analyzed for HFG and c-Met mRNA expression using RT-PCR. In addition, the binding activity of transcription factors engaged in the modulation of hepatocyte proliferation (STAT3, NF-κB) and liver-specific gene expression (C/EBP, HNF1, HNF3) was studied using an electrophoretic mobility shift assay (EMSA).

Compared to sham-BAL controls, the BAL-treated rats lived significantly longer, had better coagulation profile, higher serum glucose levels, and lower serum activity of alkaline phosphatase (Table 7.7). In BAL-treated rats, plasma HGF levels remained elevated, whereas plasma TGF-β1 levels decreased. The livers of sham-BAL rats showed no c-Met mRNA expression, whereas those of BAL-treated rats were c-Met positive. Interestingly, the core body temperature at 12 hours post-induction of FHF was significantly higher in BAL-treated rats than in sham-BAL controls.

In summary, temporary support of FHF rats with hepatocyte-based BAL prolonged survival time, improved blood coagulation, accelerated expression of c-Met HGF receptor in the native liver and lowered blood levels of a potent hepatic growth inhibitor (TGF-β1). In addition, BAL treatment resulted in increased binding of transcription factors engaged in the modulation of both hepatocyte proliferation and liver-specific gene expression. We thus demonstrated for the first time that in animals with experimentally induced FHF, hepatocyte-based therapy can not only provide metabolic support by increasing the available functional liver mass but also has a beneficial effect on the molecular biological mechanisms which control both the function and repair of the diseased liver.

Table 7.7. Laboratory values in fulminant hepatic failure rats treated with bioartificial liver loaded with microcarrier-attached syngeneic hepatocytes

Group	Survival	TGF-b1	PTT	PT	Glucose	Alk Phos	Rectal Temp
	(hr)	(ng/ml)	(sec)	(sec)	(mg/dl)	(U/l)	(°C)
1 BAL	27.4 ± 5.3	16.3 ± 5.8	34.8 ± 4.2	21.2 ± 2.5	134.8 ± 19.7	207 ± 106	33.0 ± 0.7
2 sham-BAL	17.0 ± 1.7	25.5 ± 7.4	57.0 ± 8.1	25.3 ± 0.8	90.6 ± 39.3	570 ± 187	27.9 ± 1.8

Data are shown as means ± SD; all differences between Groups 1 and 2 were at p=0.05 or less.

References

1. Kiley JE, Welch HF, Pender JC et al. Removal of blood ammonia by hemodialysis. Proc Soc Exp Biol Med 1956; 91:489-490.
2. Schechter DC, Nealon TF, Gibbon JH. A simple extracorporeal device for reducing elevated blood ammonia levels. Surgery 1958; 44:892-897.
3. Yatzidis H. The charcoal artificial kidney in clinical practice. Paper presented at Cleveland Clinic Foundation, Cleveland, Ohio, October 18, 1966.
4. Chang TMS. Hemoperfusion over microencapsulated adsorbent in a patient with hepatic coma. Lancet 1972; 2:1371.
5. O'Grady JG, Gimson AES, O'Brien CJ et al. Controlled trials of charcoal hemoperfusion and prognostic factors in fulminant hepatic failure. Gastroenterology 1988; 94:1186-1192.
6. Sorrentino F. Prime ricerche per la realizzatione di un fegato artificiale. Chir Patol Sper 1956; 4:1401-1414.
7. Hori M. Artificial liver: The concept and working hypothesis of hybrid organs: From a 25-year-old anecdote to the 21st century model. Trans AM Soc Artif Intern Organs 1982; 28:639-641.
8. Kimoto S. The artificial liver. Experimental and clinical application. Trans Am Soc Artif Intern Organs 1959; 5:102-112.
9. Mikami J, Moto M, Nishimira A et al. Surgical treatment of acute liver failure II: An experimental study of extracorporeal metabolism in the artificial liver using slices of canine liver. Jpn J Gastroenterol 1959; 56:1022.
10. Nose Y, Mikami J, Kasai Y et al. An experimental artificial liver utilizing extracorporeal metabolism with sliced and granulated canine liver. ASAIO Trans 1963; 358-362.
11. Matsumura KN, Guevara GR, Huston H et al. Hybrid bioartificial liver in hepatic failure: Preliminary clinical report. Surgery 1987; 101:151-157.
12. Eiseman B, Norton L, Kralios N.C. Hepatocyte perfusion within a centrifuge. Surg Gynecol Obst 1976; 142:21-28.
13. Olumide F, Eliashiv A, Kralios N et al. Hepatic support with hepatocyte suspensions in a permeable membrane dialyzer. Surgery 1977; 82:599-606.
14. Wolf CFW, Munkelt BF. Bilirubin conjugation by an artificial liver composed of cultured cells and synthetic capillaries. Trans Am Soc Artif Intern Organs 1975; 21:16-27.
15. Soyer T, Lempinen M, Bielanski E et al. Experimental extracorporeal hepatic assist using liver slices and cell suspension. Surg Forum 1972; 23:346.
16. Soyer T, Lempinen M, Walker JE et al. Extracorporeal assist of anhepatic animals with liver slice perfusion. Am J Surg 1973; 126:20-24.
17. Kawarabata J, Yoshida M, Suzuki T et al. Studies on extracorporeal liver support. I. Preliminary studies using isolated rabbit liver cells. Jpn J Artif Organs 1977; 6:468-471.

18. Hager JC, Carman R, Stoller R et al. A prototype for a hybrid artificial liver. Trans Am Soc Artif Intern Organs 1978; 24:250-253.
19. Takahashi I, Otsubo O, Inoue TY et al. Hepatic support system with isolated liver cells. Artif Organs 1981; 10:537-540.
20. Wada G, Oshima N. Improvement of the performance of a hybrid-type artificial liver utilizing isolated hepatocytes: The use of conditioned medium. Jpn J Artif Organs 1985; 14:241-244.
21. Kimura K, Gundermann KJ, Lie TS. Hemoperfusion over small liver pieces for liver support. Artif Organs 1980; 4:297-301.
22. Lie TS, Jung V, Kachel F et al. Successful treatment of hepatic coma by a new artificial liver device in the pig. Res Exp Med 1985; 185:483-494.
23. Kawamura A, Takahashi T, Kusumoto K et al. The development of a new hybrid support system using frozen liver pieces (FLP). Jpn J Artif Organs 1985; 14:253-257.
24. Demetriou AA, Whiting J, Levenson AM et al. New method of hepatocyte transplantation and extracorporeal liver support. Ann Surg 1986; 204:259-271.
25. Kasai S. What is needed for an artificial liver? In: Nose Y, Kjellstrand C, and Ivanovich P eds. Progress in artificial organs, Cleveland: ISAO Press 1986:711-714.
26. Kasai S, Ohno M, Asakawa M et al. Studies on hybrid artificial liver-metabolic activity of isolated hepatocytes after cryopreservation. Jpn J Artif Organs 1980; 9:390-393.
27. Saito S, Sakagami K, Orita K. A new hybrid artificial liver using a combination of hepatocytes and biomatrix. Trans Am Soc Artif Intern Organs 1987; 33:459-462.
28. Uchino J, Tsuburaya T, Kumagai F et al. 1988. A hybrid bioartificial liver composed of multiplated hepatocyte monolayers. ASAIO Trans 1988; 34:972-977.
29. Yanagi K, Ookawa K, Mizuno S et al. Performance of a new hybrid artificial liver support system using hepatocytes entrapped within a hydrogel. Trans Am Soc Artif Intern Organs 1989; 35:570-572.
30. Margulis MS, Erukhimov EA, Andreiman LA et al. Temporary organ substitution by hemoperfusion through suspension of active donor hepatocytes in a total complex of intensive therapy in patients with acute hepatic insufficiency. Resuscitation 1989; 18:85-94.
31. Jauregui HO, Santangini H, Naik S. In vitro benzodiazepine metabolism of adult rat hepatocytes seeded in hollow-fiber membrane device. Hepatology 1988; 8:1387.
32. Jauregui HO, Gann KL. Mammalian hepatocytes as a foundation for treatment in human liver failure. J Cell Biochem 1991; 45:359-365.
33. Arnaout WS, Moscioni AD, Barbour RL et al. Development of bioartificial liver: Bilirubin conjugation in Gunn rats. J Surg Res 1990; 48:379-382.
34. Shnyra A, Bocharov A, Bochkova N et al. Bioartificial liver using hepatocytes on Biosilon microcarriers: Treatment of chemically induced acute hepatic failure in rats. Artif Organs 1991; 15:189-197.
35. Takahashi M, Matsue H, Matsushita M et al. Does a porcine hepatocyte hybrid artificial liver prolong the survival time of anhepatic rabbits ? ASAIO Journal 1992; 38:M468-M472.
36. Nyberg SL, Shatford RA, Peshwa MV et al. Evaluation of a hepatocyte-entrapment hollow-fiber bioreactor: A potential bioartificial liver. Biotechnol Bioeng 1993; 41:194-203.
37. Nyberg SL, Mann HJ, Remmel RP et al. Pharmacokinetic analysis verifies P450 function during in vitro and in vivo application of a bioartificial liver. ASAIO Journal 1993; 39:M252-M256.
38. Nyberg SL, Payne WD, Amiot B, et al. Demonstration of biochemical function by extracorporeal xenohepatocytes in an anhepatic animal model. Transpl Proc 1993; 25:1944-1945.
39. Nyberg SL, Shirabe K, Peshwa MV et al. Extracorporeal application of a gel-entrapped, bioartificial liver: Demonstration of drug metabolism and other biochemical functions. Cell Transpl 1993; 2:441-452.
40. Sussman NL, Kelly JH. Improved liver function following treatment with an extracorporeal liver assist device. Artif Organs 1993; 17:27-30.
41. Sussman NL, Chong MG, Koussayir T, He DE, Shong TA, Whisennand HH, Kelly JH. Reversal of fulminant hepatic failure using an extracorporeal liver assist device. Hepatology 1992; 16:60-65.

42. Sussman NL, Gislason GT, Conlin CA et al. Use of the Hepatix extracorporeal liver assist device in the treatment of fulminant hepatic failure. Hepatology 1993; 18:65A.
43. Fremond B, Malandain C, Guyomard C et al. Correction of bilirubin conjugation in the Gunn rat using hepatocytes immobilized in alginate gel beads as an extracorporeal bioartificial liver. Cell Transpl 1993; 2:453-460.
44. Li AP, Barker G, Beck D et al. Culturing of primary hepatocytes as entrapped aggregates in a packed bed bioreactor: A potential bioartificial liver. In Vitro Cell Dev Biol 1993; 29A:249-254.
45. Neuzil DF, Rozga J, Moscioni AD et al. Use of a novel bioartificial liver in a patient with acute liver insufficiency. Surgery 1993; 113:340-343.
46. Rozga J, Williams F, Ro M-S et al. Development of a bioartificial liver: Properties and function of a hollow-fiber module inoculated with liver cells. Hepatology 1993; 17:258-265.
47. Rozga J, Holzman MD, Ro M-S et al. Hybrid bioartificial liver support treatment of animals with severe ischemic liver failure. Ann Surg 1993; 217:502-511.
48. Rozga J, Podesta L, LePage E et al. Control of cerebral oedema by total hepatectomy and extracorporeal liver support in fulminant hepatic failure. Lancet 1993; 342:898-899.
49. Takahashi T, Malchesky PS, Nose Y. Artificial liver. State of the art. Dig Dis Sc 1991; 36:1327-1340.
50. Berry MN, Friend DS. High yield preparation of isolated rat liver parenchymal cells. A biochemical and fine structural study. J Cell Biol 1969; 43:466-468.
51. Wang S, Renaud G, Infante J et al. Isolation of rat hepatocytes with EDTA and their metabolic function in primary culture. In Vitro Cell Dev Biol 1985; 21:526-530.
52. Kraemer BL, Staecker JL, Sawada N et al. Use of a low-speed, iso-density percoll centrifugation method to increase the viability of isolated rat hepatocyte preparations. In Vitro Cell Dev Biol 1986; 22:201-211.
53. Flendrig LM, Di Florio E, Ceriello A et al. In vivo investigation of a novel bioartificial liver based on porcine hepatocytes. Hepatology 1997; 26:168A.
54. Wang G, Strain AJ, Zhang W et al. Long-term culture of primary rat hepatocytes in a packed-bed bioreactor using fibra-cell discs. Hepatology 1997; 26:167A.
55. Suda J, Mochida S, Inao M et al. Hydrophilic surface coated with polyacrylamide as a tool for bioartificial liver support system: Evaluation with cultured rat hepatocytes. Hepatology 1997; 26:168A.
56. Gerlach J, Busse B, Muller C et al. Extracorporeal hybrid liver support system: Bridging to liver transplantation (OLT) in a 45 year old female with drug induced fulminant hepatic failure (FHF). Int J Artif Organs 1996; 19:544.
57. Ekevall E, Airdrie I, Cousins RB et al. Effect of oxygen partial pressure within a bioartificial liver. Int J Artif Organs 1996; 19:545.
58. Qiang S, Mitteregger R, Falkenhagen D. A novel configuration of bioartificial liver support system based on circulating microcarrier culture. Int J Artif Organs 1996; 19:547.
59. Mazariegos GV, Block GD, Khanna A et al. Preliminary canine studies of a new hollow-fiber based bioartificial liver system. ASAIO J 1997; 43:18.
60. Sajiki T, Park Y-G, Iwata H et al. Evaluation of a bioartificial liver system under in vitro perfusion culture. ASAIO J 1997; 43:16.
61. Gerlach J, Trostv T, Ryan CJ et al. Hybrid liver support system in a short term application on hepatectomized pigs. Int J Artif Organs 1994; 17:549-553.
62. Chen SC, Hewitt WR, Eguchi S et al. Prolonged survival in anhepatic pigs treated with a bioartificial liver. Surg Forum 1996; 47:161-163.
63. Ellis AJ, Hughes RD, Wendon JA et al. Pilot-controlled trial of the extracorporeal liver assist device in acute liver failure. Hepatology 1996; 24:1446-1451.
64. Watanabe FD, Mullon CJ-P, Hewitt WR et al. Clinical experience with a bioartificial liver in the treatment of severe liver failure. A phase I clinical trial. Ann Surg 1997; 225:484-494.
65. Eguchi S, Kamlot A, Ljubimova J et al. Fulminant hepatic failure in rats: Survival and effects on blood chemistry and liver regeneration. Hepatology 1996; 24:1452-1459.
66. Eguchi S, Lilja MH, Kamlot A et al. Loss and recovery of liver regeneration in rats with fulminant hepatic failure. J Surg Res 1997; 72:112-122.

CHAPTER 8

Clinical Experience with a Bioartificial Liver

Frederick D. Watanabe, Elaine Kahaku, Theodore Khalili, Paul Ting, Anthony Navarro, Achilles A. Demetriou

In spite of substantial advances in general supportive therapy and critical care, mortality in fulminant hepatic failure (FHF) remains unacceptably high, due primarily to incomplete understanding of the pathophysiology of the disease.[1] Despite this, clinicians have attempted to develop rational novel therapeutic modalities focusing mostly on plasma detoxification.[2-7] Whole liver perfusion using either human or xenogeneic organs has been used successfully to treat patients with FHF, but is limited by lack of human livers and logistical considerations in the case of animal organ use.[8-9]

In previous sections of this monograph we presented the rationale for using intact isolated cryopreserved hepatocytes to develop a bioartificial liver (BAL). Several investigators have developed and tested a variety of hepatocyte-based systems.[10-15] We have developed a BAL containing porcine hepatocytes and have demonstrated its ability to provide detoxifying and synthetic functions in a series of in vitro and in vivo animal experiments as well as pilot clinical studies.[2,16-20] The system has been described in detail in earlier sections of this monograph as well as in a more recent publication.[21] Based on the data collected from these studies, BAL design was optimized, and we arrived at the system used in the clinical trial described here. The purpose of this section is to provide an update of the clinical experience with our BAL.

Methods

Bioartificial Liver Design

Hepatocytes

Methods of porcine hepatocyte isolation, purification, attachment to a matrix, cryopreservation and storage have been previously described.[2,16-23]

System Characteristics

The design of the current BAL system (Fig. 8.1) has evolved from a series of in vitro and in vivo experimental animal and pilot clinical studies carried out over a period of several years.[2,16-23] The system has been standardized, modified and built by Circe Biomedical, Inc., Lexington, MA (HepatAssist "2000"). It has been described in detail in previous sections.

Support of the Acutely Failing Liver, 2nd ed., edited by Achilles A. Demetriou and Frederick D. Watanabe.

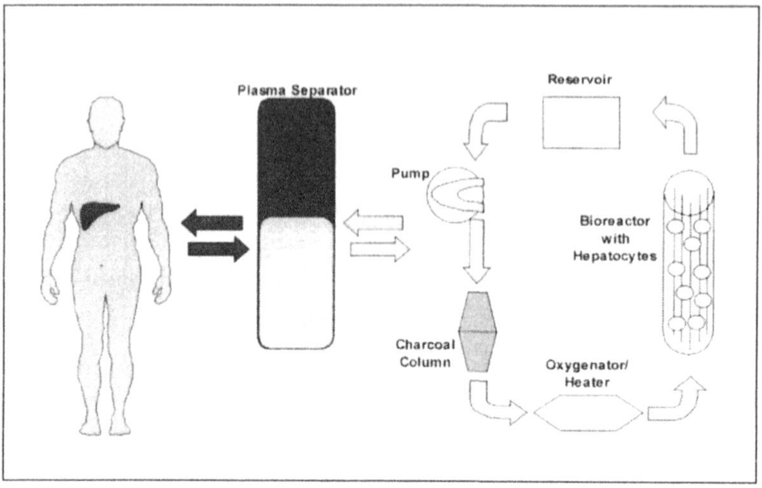

Fig. 8.1. Bioartificial liver (BAL) circuit connected to plasma separator and patient. Blood is removed from the patient and is separated into plasma and blood cells in a plasma separator. Plasma is perfused through a charcoal column to reduce its toxicity before it perfuses a bioreactor hollow-fiber module containing 5-7 billion matrix-anchored viable porcine hepatocytes. A temperature regulator and oxygenator are included in the system to provide a physiologic environment for the cells. A volume reservoir allows high recirculation flow through the circuit containing the porcine hepatocytes, thus enhancing the efficiency of the system.

Clinical Study Design

Patient Groups

All procedures were conducted in full compliance with the ethical standards of the Institutional Committee for the Protection of Human Subjects, in accordance with the Helsinki Declaration of 1975. Patients belonged to one of three groups:

Group I (n = 24) patients had no previous history of chronic liver disease, fulfilled all diagnostic criteria of FHF and were candidates for orthotopic liver transplantation (OLT) at the time of admission.

Group II (n = 3) patients had undergone OLT and developed primary nonfunction (PNF) of the transplanted liver in the immediate postoperative period with rapid clinical deterioration.

Group III (n = 10) patients presented with acute exacerbation of an underlying chronic liver disease process and were not candidates for OLT at the time of admission.

Patients in Groups I and II were immediately evaluated for OLT and were placed on a national waiting list for urgent (UNOS Status 1) transplantation. Patients in Group III were not candidates for OLT at the time they entered the study.

Patient Demographics

The demographic characteristics of the patients are summarized in Table 8.1. Patients entered the study when they developed Stage 3 encephalopathy while receiving standard medical therapy.[24] With the exception of one patient who was Stage 3 encephalopathic, all patients in Group I were in Stage 4 encephalopathy (deep coma) and exhibited signs of

Table 8.1. Patient demographic characteristics

	Mean Age (years)	Sex (Male/Female)	Number of BAL Treatments	"Bridge Time" (hours)
Group I	34.7 ± 13.6	11/13	2.3 ± 1.2	40.6 ± 28.0
Group II	48.3 ± 19.4	1/2	1.7 ± 1.1	83.0 ± 54.6
Group III	52.1 ± 11.6	7/3	1.8 ± 1.1	89.0 ± 31.0

Table 8.2. Disease etiology

	Viral	Indeterminate	Acetaminophen	Alcoholic	Ischemic	Autoimmune	PBC
Group I	3	10	8		2	1	
Group II	1	1				1	
Group III	3	1		4		1	1

brainstem dysfunction and decerebration. All three patients in Group II were Stage 4 encephalopathic. Patients in Group III were Stage 4 encephalopathic, with the exception of two patients who were in Stage 3.

Table 8.2 summarizes the etiology of liver failure. The term "indeterminate" refers to patients who clinically appear similar to the ones with FHF of viral etiology, but no viral hepatitis or other markers can be detected.

Liver Support Unit

We have established a multidisciplinary Liver Support Unit to provide optimal, focused care to patients with severe acute liver failure. The clinical team consists of physicians, nurses, pharmacists, nutritionists, physical and respiratory therapists; various areas are represented (hepatology, transplantation, anesthesiology, infectious diseases, nephrology, neurology, neurosurgery, critical care, pulmonology). The research team consists of molecular biologists, cell culture experts, physiologists, experimental surgeons, neuroscientists and others. The research and clinical teams work closely together, thus allowing seamless transfer of technology from the laboratory to the bedside. Patients are admitted to a dedicated Surgical ICU area. In addition to utilizing a computerized on-line clinical data collection system (Care Vue 9000 Clinical Information System, Hewlett-Packard, Andover, MA), we have developed and implemented standardized diagnostic and therapeutic clinical protocols to allow meaningful and accurate data collection and analysis.

Treatment Parameters/Conditions

Immediately upon admission to the Liver Support Unit, patients undergo rapid clinical evaluation. Invasive hemodynamic monitoring (peripheral arterial catheter and pulmonary artery catheter) is instituted after rapid blood product infusion to partially correct the coagulopathy. A nasogastric tube and a bladder catheter are placed. Patients with signs of intracranial hypertension have an intracranial pressure (ICP) monitor placed at the bedside (subdural) for ICP and cerebral perfusion pressure (CPP) monitoring. Patients are endotracheally intubated when in Stage 4 encephalopathy and are placed on ventilator support. The head of the bed is elevated and the room is kept dark and quiet in the case of patients with intracranial hypertension. Gastric pH is monitored and patients are placed on ulcer prophylaxis. All standard supportive medical therapy is instituted (i.e., hyperventilation, mannitol, lactulose). Supportive measures are initiated to correct electrolyte (i.e., hypocalcemia), metabolic (i.e., hypoglycemia, lactic acidemia), respiratory, coagulation and hemodynamic abnormalities. Renal dialysis and filtration are used as needed.

Vital signs, urinary output, and central hemodynamic parameters are monitored continuously. Neurologic monitoring and assessment consisting of hourly clinical examination, serially assessing brainstem dysfunction by using the comprehensive level of consciousness score (CLOCS) system, determination of the Glasgow coma score (GCS) and continuous monitoring of ICP and CPP, are carried out in all patients with clinical evidence of brainstem dysfunction. Liver function tests, coagulation tests, complete blood cell count, serum electrolytes, ammonia, lactate, blood urea nitrogen and creatinine levels are determined at the start, mid-point and end of each BAL treatment. Arterial blood gases are determined serially.

A double lumen catheter is placed in the right superficial femoral vein for BAL treatments (and renal dialysis if needed). Each BAL treatment lasts six hours and is preceded by one hour of plasma separation (not exchange) to ascertain patient stability and ensure that patients tolerate the extracorporeal circuit. Sodium citrate is used as an anticoagulant. Plasma ionized calcium levels are measured hourly during BAL treatment to prevent ionized hypocalcemia due to the calcium-chelating effect of citrate. Calcium chloride infusion is initiated at the start of BAL treatment. Blood is removed from the patient at about 90-100 ml/min and is separated into blood cells and plasma; the latter enters the BAL circuit at 50 ml/min and then the plasma reservoir. Plasma recirculates through the BAL at 400 ml/min. Plasma is maintained at physiologic temperature and is oxygenated. Finally, plasma is reconstituted with blood cells and returns to the patient at the same rate at which it is removed (90-100 ml/min). The complete system is shown diagrammatically in Figure 8.1.

Data Analysis

Statistical analysis was carried out using the paired Student "t" test to compare immediately pre- and post-BAL treatment values. A p value < 0.05 is considered significant. Data are expressed as means ± standard errors.

Results

BAL Treatment

Patients tolerated BAL treatments well. With the exception of one patient who experienced an episode of transient hypotension requiring discontinuation of a treatment, patients remained hemodynamically stable during treatments (Table 8.3). No technical problems were identified during plasma separation and BAL perfusion (i.e., embolization, thrombosis, bleeding). All patients were supplemented with calcium chloride to prevent ionized hypocalcemia. There were no adverse hypersensitivity or immunologic reactions to the use of porcine hepatocytes. There was no evidence that treatment with the BAL had an adverse effect on subsequent liver allograft survival and function.

Table 8.3. Effect of BAL treatment on hemodynamic parameters

	Pre-BAL	Post-BAL	p <
Group I			
Heart Rate (/min)	99 ± 2	96 ± 2	0.4
MAP (mmHg)	88 ± 2	86 ± 2	0.3
CVP (cm H_2O)	8.7 ± 0.6	9.0 ± 0.6	0.6
Cardiac Index	4.9 ± 0.2	4.6 ± 0.2	0.2
Group II			
Heart Rate (/min)	102 ± 5	91 ± 3	0.1
MAP (mmHg)	79 ± 3	88 ± 6	0.2
CVP (cm H_2O)	14.0 ± 2.0	14.9 ± 3.0	0.9
Cardiac Index	6.5 ± 2.5	6.4 ± 2.6	0.7
Group III			
Heart Rate (/min)	100 ± 3	100 ± 3	1.0
MAP (mmHg)	81 ± 3	82 ± 4	0.7
CVP (cm H_2O)	12.4 ± 1.4	14.0 ± 1.3	0.2
Cardiac Index	5.8 ± 0.4	5.4 ± 0.4	0.1

BAL: Bioartificial Liver; MAP:Mean Arterial Pressure; CVP: Central Venous Pressure

Patient Survival

Group I (n = 24). All patients who were candidates for OLT were successfully "bridged" to OLT (n = 18). One patient with FHF due to ischemic liver injury following a heat stroke was initially a candidate for transplantation, underwent five treatments with the BAL and demonstrated remarkable neurologic recovery; however, he subsequently developed severe ischemic necrotizing pancreatitis, was removed from the transplant list and expired 21 days later in multiorgan failure. All 18 patients who were successfully "bridged" to OLT have experienced full neurologic recovery and were discharged from the hospital. Five patients were treated with the BAL and recovered fully without undergoing OLT.

Group II (n = 3). All three patients in this group were successfully "bridged" to re-transplantation, experienced full recovery and have been discharged from the hospital.

Group III (n = 10). Patients in this group experienced transient clinical improvement after BAL treatment. However, only two patients were able to recover enough native liver function to survive the acute-on-chronic failure; these two patients later became candidates for OLT (in one case following treatment of active sepsis and in the other after entering an alcohol rehabilitation program for six months). The remaining eight patients died 1-21 days (mean: 7.1 days) following their last BAL treatment, from variceal bleeding, sepsis and multiorgan failure.

Neurologic Effects of the BAL

Patients in Group I experienced remarkable neurologic improvement with reversal of the decerebrate state following BAL treatments(s); posturing, anisocoria, sluggish pupil reactivity were lessened and patients became more responsive to external stimuli. Brainstem function improved as shown by the higher CLOCS scores (Table 8.4). There was a significant reduction in ICP (Table 8.4). This can be better appreciated if one examines more closely patients whose ICP was high. In 25 BAL treatments in which the ICP was > 20 mm Hg, the effect of BAL treatment was dramatic (ICP: 22.7 ± 2.0 to 13.9 ± 1.3 mmHg, $p < 0.0005$; CPP: 68.7 ± 3.9 to 79.0 ± 2.8 mmHg, $p < 0.003$; CLOCS: 21.3 ± 1.1 to 27.0 ± 1.0, $p < 0.000002$).

It is difficult to quantitate the neurologic impact of BAL treatment in Group II patients because they were in the postanesthetic period under the influence of paralyzing drugs and other medications. Group III patients experienced transient neurologic improvement after BAL treatments, manifested primarily by increased responsiveness. This observation, however, was difficult to quantitate. There was an improvement in the CLOCS score due to the effect seen in a small number of patients in this group who experienced a degree of brainstem dysfunction.

Other Effects of the BAL

The effects of BAL treatment on liver function tests, metabolic and renal function are shown in Tables 8.5-8.6. A decrease in ammonia, transaminases and bilirubin levels was observed. Plasma amino acid levels were measured at the beginning and end of each BAL treatment. The ratio of branched chain amino acids (BCAA: leucine, isoleucine, valine) to

Table 8.4. Effect of BAL on neurologic parameters

	Pre-BAL	Post-BAL	p <
Group I			
ICP (mmHg)	17.0 ± 1.5	10.9 ± 1.0	0.0002
CPP (mmHg)	70 ± 2	75 ± 2	0.04
GCS	6.8 ± 0.4	7.4 ± 0.4	0.01
CLOCS	24.7 ± 1.2	32.0 ± 1.1	0.000001
Group II			
GCS	5.0 ± 1.1	7.0 ± 1.4	0.2
CLOCS	29.7 ± 7.4	31.7 ± 7.9	0.5
Group III			
ICP (mmHg)	12.3 ± 0.9	14.0 ± 1.5	0.4
CPP (mmHg)	85 ± 1	98 ± 8	0.3
GCS	8.2 ± 0.7	8.4 ± 0.7	0.4
CLOCS	29.7 ± 2.3	34.0 ± 1.7	0.001

BAL: Bioartificial Liver; ICP: Intracranial Pressure; CPP: Cerebral Perfusion Pressure; GCS: Glasgow Coma Score; CLOCS: Comprehensive Level of Consciousness Score.

Table 8.5. Effect of BAL on liver function tests

	Pre-BAL	Post-BAL	p <
Group I			
AST (U/L)	1255 ± 261	879 ± 148	0.002
ALT (U/L)	1075 ± 184	674 ± 120	0.000005
Alkaline Phosph.(U/L)	116 ± 7	90 ± 5	0.0000001
Total Bilirubin (mg/dl)	17.9 ± 1.5	14.6 ± 1.2	0.000001
Dir. Bilirubin (mg/dl)	8.5 ± 1.0	6.4 ± 0.6	0.000001
Indir. Bilirubin (mg/dl)	9.3 ± 0.8	7.9 ± 0.7	0.0000007
Group II			
AST (U/L)	5661 ± 2613	2821 ± 1291	0.1
ALT (U/L)	2139 ± 704	1633 ± 544	0.05
Alkaline Phosp. (U/L)	108 ± 15	83 ± 9	0.03
Total Bilirubin (mg/dl)	19.1 ± 2.2	14.7 ± 1.7	0.009
Dir. Bilirubin (mg/dl)	3.8 ± 1.3	3.1 ± 0.8	0.2
Indir. Bilirubin (mg/dl)	15.3 ± 2.3	11.6 ± 1.3	0.05
Group III			
AST (U/L)	692 ± 374	723 ± 409	0.5
ALT (U/L)	349 ± 126	281 ± 114	0.06
Alkaline Phosph. (U/L)	117 ± 10	119 ± 22	0.9
Total Bilirubin (mg/dl)	26.0 ± 2.7	21.6 ± 2.2	0.000003
Dir. Bilirubin (mg/dl)	12.8 ± 1.5	10.3 ± 1.2	0.000006
Indir. Bilirubin (mg/dl)	13.2 ± 1.8	11.3 ± 1.5	0.002

BAL: Bioartificial Liver; AST: Aspartate Aminotransferase; ALT: Alanine Aminotransferase

aromatic amino acids (AAA: phenylalanine, tyrosine), felt to be an index of the degree of encephalopathy, was calculated.[25] There was a significant ($p < 0.01$) increase (improvement) in the BCAA/AAA ratio from 0.75 ± 0.07 to 0.98 ± 0.07. The increase was primarily due to reduction in AAA levels. No improvement in prothrombin time and other coagulation parameters was seen, but most of the patients had already been treated with fresh frozen plasma and clotting factor concentrates.

Table 8.6. Effect of BAL on metabolic and renal function parameters

	Pre-BAL	Post-BAL	p <
Group I			
Glucose (mg/dL)	126 ± 5	175 ± 7	0.0000006
Ammonia (mmol/L)	160 ± 8	134 ± 6	0.0002
Lactate (mmol/L)	4.4 ± 0.7	4.2 ± 0.6	0.2
Albumin (g/dl)	3.12 ± 0.08	2.6 ± 0.1	0.0000006
BUN (mg/dl)	18.9 ± 2.5	17.0 ± 2.0	0.0000008
Creatinine (mg/dl)	1.5 ± 0.2	1.1 ± 0.1	0.000001
Group II			
Glucose (mg/dl)	117 ± 26	144 ± 24	0.06
Ammonia (m ol/L)	81 ± 9	91 ± 13	0.03
Lactate (mmol/L)	13.1 ± 2.9	13.2 ± 2.2	0.9
Albumin (g/dl)	3.7 ± 0.3	2.7 ± 0.1	0.01
BUN (mg/dl)	12.0 ± 2.4	11.5 ± 2.7	0.4
Creatinine (mg/dl)	1.6 ± 0.3	1.6 ± 0.3	1.0
Group III			
Glucose (mg/dl)	141 ± 9	171 ± 11	0.001
Ammonia (mmol/L)	173 ± 31	131 ± 15	0.08
Lactate (mmol/L)	5.7 ± 1.1	5.6 ± 0.9	0.9
Albumin (g/dl)	3.0 ± 0.1	2.6 ± 0.1	0.00003
BUN (mg/dl)	12.8 ± 1.5	35.5 ± 3.2	0.0004
Creatinine (mg/dl)	2.8 ± 0.3	2.2 ± 0.2	0.00002

BAL: Bioartificial Liver; BUN: Blood Urea Nitrogen

Discussion

Multiple attempts have been made to develop extracorporeal and other systems to provide support for the failing liver.[3-21] The strategies and the pros and cons of the various systems have been previously discussed.[2,22] The complexity of the liver is such that we are unable to provide full replacement of all liver functions for any significant length of time with any of the existing systems, including ex vivo perfusion with intact human liver. The question then becomes: What are the relatively few critical liver functions needed to improve patient survival?

It is clear that in the case of FHF patients, for a liver support system to be effective, it has to arrest and/or reverse the rapid development of intracranial hypertension leading to brainstem herniation and death. If a system can achieve that, it will have a significant impact on outcome in this group of patients. Our data suggest that state of the art critical care in addition to the BAL can result in significant reduction in intracranial pressure. It is unlikely that BAL treatment alone, in the absence of aggressive standard medical therapy, can arrest/reverse cerebral edema. At the same time, excellent medical therapy alone has historically not been successful in preventing permanent neurologic damage. OLT is the only relatively novel therapeutic modality which has had a significant impact on outcome in FHF patients.[1] Therefore, a system which can arrest intracranial hypertension as a "bridge" to OLT, will ultimately result in improved patient survival. The BAL, based on the preliminary data presented here, appears to be an effective "bridge" to OLT. Further appropriately controlled trials have been initiated to confirm this observation. Certainly the ultimate goal of a liver support treatment is to treat patients early enough and repeatedly, to allow patients to regenerate their livers and recover normal function. Five patients in Group I fell in this category and had full recovery without a transplant.

Use of the BAL in PNF patients is justified only if the system can provide effective metabolic and physiologic support. The small number of patients treated with the BAL does not, at this stage, allow any meaningful analysis and conclusions. These patients need to be studied prospectively in a controlled trial. Assessing the efficacy of any support system in these patients is difficult. The presence of anesthetics and paralyzing agents makes neurologic assessment of the patients difficult. Rendering these patients, as well as patients with massive liver necrosis, anhepatic, may make it possible to provide more effective support with the BAL. These issues need, however, to be tested and resolved in the experimental laboratory.

We have carried out a phase I clinical trial to examine primarily the safety of the BAL. The remarkable survival obtained in the group of patients with FHF is due to a combination of excellent clinical team skills and possibly the introduction of this innovative liver support technology. The role of the latter remains to be better defined in a controlled randomized trial setting. For this or other liver support systems to play a meaningful role in the management of patients with exacerbations of underlying liver disease processes, such patients need to be treated early, while they have a residual liver mass which can potentially recover following the acute precipitating event. These patients will also need to be treated for longer periods of time.

References

1. Lee WM. Acute liver failure. N Eng J Med 1993;329:1862-1868.
2. Demetriou AA. Support of the acutely failing liver. Medical Intelligence Unit. Austin: RG Landes Co., 1994.
3. Bihari D, Hughes RD, Gimson AES et al. Effect of serial resin hemoperfusion in fulminant hepatic failure. Int J Artif Organs 1983; 6:299-302.
4. Gimson AES, Mellon PJ, Braude S et al. Earlier charcoal hemoperfusion in fulminant hepatic failure. Lancet 1982; ii:68-83.
5. Silk DBA, Trewby PN, Chase RA et al. Treatment of fulminant hepatic failure by polyacrylonitrile membrane hemodialysis. Lancet 1977; ii:1-3.
6. Redeker AG, Yamahiro HS. Controlled trial of exchange transfusion therapy in fulminant hepatitis. Lancet 1973; ii:3-6.
7. Cooper GN, Karlson KE, Clowes GHA et al. Total body washout and exchange. A valuable tool in acute hepatic coma and Reye's syndrome. Am J Surg 1977; 133:522-530.

8. Fox IJ, Langnas AN, Ozaki CF et al. Successful application of extracorporeal liver perfusion for the treatment of fulminant hepatic failure; a technology whose time has come. Am J Gastroenterol 1993; 88:1876-1881.
9. Chari RS, Collins BH, Maee JC et al. Brief report: Treatment of hepatic failure with ex vivo pig liver perfusion followed by liver transplantation. N Eng J Med 1994; 331:234-237.
10. Olumide F, Eliashiv A, Kralios N et al. Hepatic support with hepatocyte suspensions in a permeable membrane dialyzer. Surgery 1977; 82:599-606.
11. Matsumura KN, Guevara GR, Houston H et al. Hybrid bioartificial liver in hepatic failure: Preliminary clinical report. Surgery 1987; 101:99-103.
12. Saito S Sakagami K, Orita K. A new hybrid artificial liver using a combination of hepatocytes and biomatrix. Trans Am Soc Artif Organs 1987; 33:459-462.
13. Jauregui HO, Naik S, Driscoll J et al. Adult rat hepatocyte cultures as the cellular component of an artificial hybrid liver. In: Paul JP, Galor JDS, Courtney JM, Gilchrist T, eds. Biomaterials in artificial organs. London: Macmillan,1984:130-140.
14. Nyberg SL, Remmel PR, Mann HJ et al. Primary hepatocytes outperform HepG2 cells as the source of biotransformation functions in a bioartificial liver. Ann Surg 1994; 220:59-67.
15. Sussman NL, Chong MG. Koussayir T et al. Reversal of fulminant hepatic failure using an extracorporeal liver assist device. Hepatology 1992; 16:60-65.
16. Neuzil DF, Rozga J, Moscioni AD et al. Use of a novel bioartificial liver in a patient with acute liver insufficiency. Surgery 1993; 113:340-343.
17. Rozga J, Holzman MD, Ro MS et al. Hybrid bioartificial liver support treatment of animals with severe ischemic liver failure. Ann Surg 1993; 217:502-511.
18. Rozga J, Williams F, Ro MS et al. Development of a bioartificial liver: Properties and function of a hollow-fiber module inoculated with liver cells. Hepatology 1993; 17:256-265.
19. Demetriou AA, Rozga J, Podesta L et al. Early clinical experience with a bioartificial liver. Scand J Gastroenterol 1995; 30 Suppl 208:111-117.
20. Rozga J, LePage E, Moscioni AD et al. Clinical use of a bioartificial liver to treat fulminant hepatic failure. Ann Surg 1994; 219:538-546.
21. Watanabe FD, Mullon CJP, Hewitt WR et al. Clinical experience with a bioartificial liver in the treatment of severe liver failure: A phase I clinical trial. Ann Surg 1997; 225:484-494.
22. Kamlot A, Rozga J, Watanabe FD, Demetriou AA. Review: Artificial liver support systems. Biotech Bioeng 1996; 50:382-391.
23. Morsiani E, Rozga J, Scott HC et al. Automated liver cell processing facilitates large-scale isolation and purification of porcine hepatocytes. ASAIO J 1995; 41:155-161.
24. Trey C, Davidson C. The management of fulminant hepatic failure. In: Popper H, Schaffner F, eds. Progress in liver disease New York: Grune & Stratton, 1970; 282-298.
25. Record CO, Buxton B, Chase R et al. Plasma and brain amino acids in fulminant hepatic failure and their relationship to hepatic encephalopathy. Eur J Clin Invest 1976; 6:387-394.

CHAPTER 9

Clinical Use of a Bioartificial Liver to Treat Acetaminophen-Induced Fulminant Hepatic Failure

Olivier Detry, Nikolaos Arkadopoulos, Elaine Kahaku, Frederick D. Watanabe, Jacek Rozga, Achilles A. Demetriou

Acetaminophen toxicity may cause massive hepatocellular necrosis leading to fulminant hepatic failure (FHF).[1,2] In acetaminophen-induced FHF patients, prognosis can be accurately assessed using criteria defined by the King's College Hospital (KCH) group.[3] According to these widely accepted and validated criteria, in the absence of orthotopic liver transplantation (OLT) the probability of patient death is 90% if:

1. The arterial blood pH is less than 7.3 at 24 hours or later after acetaminophen ingestion, assuming appropriate volume loading; or
2. A serum creatinine level greater than 300 µmol/l, hepatic encephalopathy grade of III-IV and an international normalized ratio (INR) greater than 7 are concurrently present.[4-7]

OLT remains the only definitive treatment for FHF patients who fulfill the above criteria. However, access to OLT is limited by organ donor shortage, psychiatric and other contraindications as well as high cost and significant associated morbidity.[5,8] In our institution, we have developed a treatment strategy based on the use of an extracorporeal bioartificial liver (BAL) support system.[9] In this section, we describe the successful clinical application of BAL support to treat patients with acetaminophen-induced FHF.

Methods

All procedures were conducted in full compliance with the standards of the Institutional Committee for the Protection of Human Subjects, in accordance with the Helsinki Declaration of 1975. Eight patients with acetaminophen-induced FHF who fulfilled the KCH criteria were treated in our Liver Support Unit (Table 9.1). All patients were admitted to a dedicated surgical intensive care unit and were evaluated for urgent OLT. All patients were endotracheally intubated for airway protection and were treated with standard support measures to correct metabolic, respiratory and hemodynamic abnormalities. Invasive hemodynamic monitoring was instituted after partial correction of the coagulopathy. N-Acetylcysteine was administered through the nasogastric tube (loading dose 140 mg/kg, then 70 mg/kg every 4 hours for 17 total doses). Fresh frozen plasma was infused as needed. Vital signs, urinary output and hemodynamic parameters were continuously recorded (CareVue 9000 Clinical Information System; Hewlett-Packard, Andover, MA). Brain stem dysfunction was serially assessed by using the Glasgow coma and comprehensive level of

Support of the Acutely Failing Liver, 2nd ed., edited by Achilles A. Demetriou and Frederick D. Watanabe.

Table 9.1. Demographics, prognostic factors and outcome of patients with acetaminophen-induced fulminant hepatic failure

Case	Age (yr)	Gender	pH	INR	Creatinine (mmol/l)	Encephalopathy Grade	Highest ICP (mmHg)	OLT Candidate	No of BAL Treatments	Outcome
1	18	Female	7.07	11.3	115	4	32	yes	3	OLT
2	50	Female	7.13	7.0	212	4	52	yes	2	OLT
3	47	Female	7.10	4.8	476	4	12	yes	2	Recovery
4	19	Female	6.96	9.3	229	4	49	yes	2	OLT
5	34	Female	7.05	12.9	238	3	NA	no	4	Recovery
6	35	Male	7.23	5.1	88	4	51	no	2	Recovery
7	37	Female	7.13	4.0	273	3	NA	no	2	Recovery
8	27	Female	7.15	15.7	203	4	45	no	4	Recovery

Abbreviations: BAL, bioartificial liver; ICP, intracranial pressure; NA, not available; OLT, orthotopic liver transplantation.

consciousness score (CLOCS) score systems.[10] A subdural fiberoptic transducer (Model 110-4B; Camino Labs, San Diego, CA) was inserted in FHF patients in stage 4 encephalopathy; intracranial pressure (ICP) and cerebral perfusion pressure (CPP) were continuously recorded. CPP was calculated as the mean arterial pressure minus ICP. Medical treatment for intracranial hypertension included fluid restriction and hyperventilation (PCO_2 between 25 and 30 mmHg). The goal was to maintain the ICP below 20 mmHg and the CPP above 50 mmHg. Intracranial hypertension refractory to hyperventilation was treated in the following sequence:

1. Intravenous bolus of mannitol (0.5 to 1 gm/kg every 6 hours if needed and if blood osmolality < 310 mosmol/l);
2. Intravenous fentanyl bolus (0.25-1 mg); or
3. Intravenous pentobarbital (bolus of 3 to 5 mg/kg). Patients with elevated ICP were maintained in a quiet, dark room with their head and chest elevated at a 30° angle.

All patients were treated with a bioartificial liver (BAL) daily until OLT, recovery or death.[9] The BAL system (HepatAssist 2000; Circe Biomedical, Lexington, MA) has been described previously. Various parameters were measured. Liver function tests, coagulation tests, ammonia, lactic acid, blood urea nitrogen and creatinine levels were determined at the start and end of each BAL treatment.

Data are expressed as means ± SEM. Statistical analysis was carried out using the paired Student's t-test to compare immediately pre- and post-BAL treatment values. Yates' modification of Chi-squared test for small populations was used to compare the survival rate of the nontransplanted patients to expected survival according to the KCH prognostic criteria. A p value < 0.05 was considered statistically significant.

Results

All patients tolerated BAL treatments without clinically significant adverse effects (Table 9.2). Three patients were transplanted as soon as a liver graft became available, and recovered. Five patients recovered after 2-4 BAL treatments (Table 9.1). All of the eight acetaminophen-induced FHF patients enjoyed full recovery without neurological sequelae and were discharged from the hospital. The survival rate of nontransplanted patients (n = 5) was significantly higher than that which would be expected according to the KCH criteria ($p < 0.05$). Patients experienced neurologic improvement after BAL treatment, as demonstrated by improvement of CLOCS and significant reduction in ICP (Table 9.3). The effect of BAL treatment on liver function, metabolic parameters and renal function is shown in

Table 9.2. Effect of BAL treatment on hemodynamic parameters

	Pre-BAL	Post-BAL	p
Heart Rate (/min)	107 ± 3	104 ± 4	NS
MAP (mmHg)	91 ± 2	89 ± 3	NS
CVP (mmHg)	8 ± 1	8 ± 1	NS
PCWP (mmHg)	15 ± 2	11 ± 1	NS
SVRI (dyne sec cm^{-5} m^2)	1241 ± 145	1586 ± 257	NS
CI (l/min m^2)	5.4 ± 0.4	4.5 ± 0.4	NS

Abbreviations: BAL, bioartificial liver; CI, cardiac index; CVP, central venous pressure; MAP, mean arterial pressure; NS, not significant; PCWP, pulmonary capillary wedge pressure; SVRI, systemic vascular resistance index.

Table 9.4. A significant decrease in ammonia and transaminase levels and an improvement in renal function were also observed.

Discussion

Acetaminophen toxicity is a common cause of acute liver failure.[1,11] Intensive care management in dedicated liver units, combined with OLT, has resulted in a significant improvement in overall survival.[11] Nevertheless, patients who fulfill the KCH criteria still face dismal prognosis if OLT is not performed in a timely manner. Classically, management of these patients includes listing for urgent OLT if no contraindications are present and conservative measures aimed at controlling brain edema (hyperventilation, mannitol, barbiturates, etc.) until a liver graft is procured.[1,6,7] However, this approach has yielded disappointing results primarily for two reasons:

Table 9.3. Effect of BAL on neurologic parameters

	Pre-BAL	Post-BAL	p
ICP (mmHg)	20.5 ± 3.1	15.5 ± 1.8	0.04
CPP (mmHg)	71.7 ± 4.0	76.4 ± 2.9	NS
GCS	7.0 ± 0.6	7.5 ± 0.6	NS
CLOCS	28.0 ± 2.2	32.6 ± 2.0	0.0006

Abbreviations: BAL, bioartificial liver; CLOCS, comprehensive level of consciousness score; CPP, cerebral perfusion pressure; GCS, Glasgow coma scale; ICP, intracranial pressure; NS, not significant.

Table 9.4. Effect of BAL on liver and renal function, and metabolic parameters

	Pre-BAL	Post-BAL	p
PT (sec)	21 ± 1	23 ± 1	0.05
AST (U/l)	2522 ± 579	1627 ± 325	0.003
ALT (U/l)	1929 ± 291	1214 ± 217	0.00005
Conjugated Bilirubin (mmol/l)	58 ± 7	48 ± 5	0.002
Unconjugated Bilirubin (mmol/l)	52 ± 3	51 ± 5	NS
Lactate (mmol/l)	4.1 ± 0.6	4.2 ± 0.5	NS
Ammonia (mmol/l)	169 ± 14	137 ± 13	0.01
BUN (mmol/l)	7.9 ± 1.5	6.8 ± 1.4	0.0003
Creatinine (mmol/l)	133 ± 27	97 ± 18	0.009

Abbreviations: ALT, alanine aminotransferase; AST, aspartate aminotransferase; BAL, bioartificial liver; BUN, blood urea nitrogen; NS, not significant; PT, prothrombin time.

1. The majority of patients who fulfill the KCH criteria are not listed for OLT because of medical or psychosocial contraindications. Makin et al reported that 55% of patients with acetaminophen-induced FHF had developed medical contraindications to OLT at the time of admission.[6] In addition, suicidal behavior and weak psychosocial support are common findings in this group of patients; such psychiatric considerations may become even stronger contraindications to OLT as the organ shortage worsens.[5,12]
2. A significant percentage (30%-50%) of patients who are initially listed for urgent OLT develop irreversible brain damage or die before a donor organ becomes available.[4,6,13]

In our series, all patients met the KCH criteria by having significant metabolic acidosis (arterial pH < 7.3), which has been shown to be the best single predictor of mortality after acetaminophen overdose.[3,5-7,11,14,15] In addition, 6 of 8 patients were in encephalopathy stage 4 and 5 of 8 had severe intracranial hypertension. All patients demonstrated significant neurologic and metabolic improvements following BAL treatment(s). Three patients were successfully bridged to OLT at 20-58 hours after listing. The most interesting effects of the BAL treatments were shown in the five nontransplanted patients who fully recovered, despite a predicted spontaneous mortality of 90%. OLT remains the gold standard of treatment for FHF patients who meet appropriate criteria. Therefore, as a rule, BAL treatments were discontinued as soon as a liver donor became available.

It is still unclear which component of our treatment strategy is responsible for the outstanding survival results. Medical measures alone in patients who fulfill the KCH criteria have consistently resulted in a mortality rate of > 85%, even in the most recent clinical reports by Bismuth et al and by Anand et al.[5,7] Furthermore, in these patients, charcoal hemoperfusion either alone or combined with high-flux dialysis has failed to produce a significant survival benefit.[16,17] Based on extensive experimental and clinical data, we speculate that the BAL made a contribution in the observed beneficial effects.[9,18,19]

In conclusion, BAL appears to improve the outcome of high-risk patients with acetaminophen-induced FHF, even in the absence of OLT. Avoiding OLT is particularly important in an era of growing concern about organ shortage, high cost of transplants and need for life-long immunosuppression. A randomized, controlled, prospective study is now initiated to determine if early (encephalopathy stage 1-2) BAL treatment can further improve the chance of spontaneous recovery of patients with acetaminophen poisoning.

References

1. Hoofnagle JH, Carithers RL, Shapiro C et al. Fulminant hepatic failure: Summary of a workshop. Hepatology 1995; 21:240-252.
2. Trey C, Davidson C. The management of fulminant hepatic failure. Prog Liv Dis 1970; 3:282-298.
3. O'Grady JG, Alexander GJ, Hayllar KM et al. Early indicators of prognosis in fulminant hepatic failure. Gastroenterology 1989; 97:439-445.
4. Mutimer DJ, Ayres RC, Neuberger JM et al. Serious paracetamol poisoning and the results of liver transplantation. Gut 1994; 35:809-814.
5. Anand AC, Nightingale P, Neuberger JM. Early indicators of prognosis in fulminant hepatic failure: An assessment of the King's criteria. J Hepatol 1997; 26:62-68.
6. Makin AJ, Wendon J, Williams R. A 7-year experience of severe acetaminophen-induced hepatotoxicity (1987-1993). Gastroenterology 1995; 109:1907-1916.
7. Bismuth H, Samuel D, Castaing D et al. Liver transplantation in Europe for patients with fulminant hepatic failure. Semin Liver Dis 1996; 16:415-425.
8. Forster J, Bartholome WG, Delcore R. Should a patient who attempted suicide receive a liver transplant? J Clin Ethics 1997; 7:257-267.

9. Watanabe FD, Mullon CJP, Hewitt WR et al. Clinical experience with a bioartificial liver in the treatment of severe liver failure: A phase I clinical trial. Ann Surg 1997; 225:484-494.
10. Stanczak DE, White JGI, Gouview WD et al. Assessment of level of consciousness following severe neurological insult. J Neurosurg 1984; 60:955-960.
11. Williams R. Classification, etiology, and considerations of outcome in acute liver failure. Semin Liver Dis 1986; 16:343-348.
12. Lidofsky SD. Fulminant hepatic failure. Crit Care Clin 1995; 11:415-430.
13. Larsen FS, Kirkegaard P, Rasmussen A et al. The Danish liver transplantation program and patients with serious acetaminophen intoxication. Transpl Proc 1995; 27:3519-3520.
14. Pauwels A, Mostefa-Kara N, Florent C et al. Emergency liver transplantation for acute liver failure: Evaluation of London and Clichy criteria. J Hepatol 1993; 17:124-127.
15. Harrison PM, Keays R, Bray GP et al. Improved outcome of paracetamol induced fulminant hepatic failure by late administration of acetylcysteine. Lancet 1990; 335: 1572-1575.
16. O'Grady JG, Gimson AE, O'Brien CJ et al. Controlled trials of charcoal hemoperfusion and prognostic factors in fulminant hepatic failure. Gastroenterology 1988; 94:1186-1193.
17. Higgins RM, Goldsmith DJ, MacDiarmid-Gordon A et al. Treating paracetamol overdose by charcoal hemoperfusion and long-hours high-flux dialysis. Q J Med 1996; 89:297-306.
18. Chen S, Hewitt W, Eguchi S et al. Prolonged survival in anhepatic pigs treated with a bioartificial liver. Surgical Forum 1996; 48:161-163.
19. Rozga J, Williams F, Ro M et al. Development of a bioartificial liver: Properties and function of a hollow-fiber module inoculated with liver cells. Hepatology 1993; 17:258-265.

CHAPTER 10

Development of a Strategy for Management of Severe Acute Liver Failure: The Liver Support Unit

Achilles A. Demetriou, Frederick D. Watanabe

As we have repeatedly emphasized throughout this monograph, severe acute liver failure (SALF) is associated with very high morbidity and mortality and represents one of the most challenging problems in clinical medicine. Historically, when faced with complex clinical problems, the medical community responded by developing specialized, multidisciplinary, focused programs to manage them. This has resulted in the development of Burn Units for managing major burns, Trauma Centers for treating severely injured patients and other such specialized Units. The rationale behind this strategy has been that managing these challenging problems requires a team approach crossing multiple discipline boundaries, commitment of substantial institutional resources, availability of a wide patient referral base and recognition that it will result in improved clinical outcomes.

Team approach is essential due to the complexity of the underlying medical problem. Commitment of resources is needed because of the high level of severity of illness, need for specialized intensive care and sophisticated invasive monitoring, utilization of multiple diagnostic procedures and a variety of major surgical interventions. A wide patient referral base is essential to provide adequate numbers of patients so that a high level of expertise can be achieved. Positive impact of the specialized Unit on disease outcome will make possible continuing commitment of resources.

Defining the Clinical Need

The most compelling argument for the establishment of a specialized Liver Support Unit (LSU) is the complex nature of the disease.[1,2] The major goal of the LSU is to improve clinical outcomes. Other goals include: provision of an environment allowing performance of meaningful controlled studies examining the impact of novel therapeutic modalities on outcome, serving as a training facility for physicians and other medical professionals entering the field and being a resource for the community. The LSU has to achieve credibility by demonstrating clinical expertise, ability to introduce innovations and demonstrating that patients enter critically developed clinical pathways which assure high quality of care and that they do not receive unproved therapies unless they participate in specific controlled clinical trials.

Liver Support Unit Logistics

The clinical leadership for the LSU team should be provided by the physician with the interest and expertise in this area. This can be a hepatologist, surgeon, intensivist or

Support of the Acutely Failing Liver, 2nd ed., edited by Achilles A. Demetriou and Frederick D. Watanabe.

anesthesiologist. The dedicated site has to be the best Critical care Unit (Surgical, Medical, etc.) with a dedicated cadre of trained Nursing Staff. Eventually, with the success of the program, such a Unit will become exclusively dedicated to the care of patients with liver failure.

The two major ingredients for the success of the LSU are:

1. The Staff of the Unit needs to have in-depth knowledge of the pathophysiology of the disease and understand the important points of departure from the usual Critical care Unit patient diagnosis and management. Patients need to be cared for under strict protocols, and members of the team have to understand how and why these protocols are being used. Nurses must be trained to recognize early signs of development of cerebral edema and neurologic deterioration.
2. A senior physician has to be at the bedside to care for patients with intracranial hypertension. This is necessary because of the requirement for rapid intervention during a narrow window of therapeutic opportunity before a patient develops irreversible brain damage.

Treatment of SALF patients requires a multidisciplinary effort. The team has to have expertise in a number of areas (adult and pediatric): hepatology, hepatobiliary surgery and liver transplantation, critical care, nursing, respiratory therapy, nutrition, pharmacy, anesthesiology, neurology and neurosurgery, pathology, nephrology, cardiology, psychiatry, social services, infectious diseases and pulmonary medicine. The above components are needed to provide optimal and responsive clinical care.

The second, equally important, component of an LSU is a team of researchers who are part of the program. This group should include hepatic physiologists, experimental surgeons, neurophysiologists, molecular physiologists, molecular pathologists, cell culture experts, detoxification/cytochrome P-450 experts, tissue and materials engineers, immunologists, analytical chemists and others. In spite of a massive body of literature on the pathophysiology of SALF, many important issues remain unresolved. The LSU provides an opportunity to use clinical observations as a starting point for meaningful basic scientific investigation. There has to be close interaction between clinicians and basic scientists to carry out well-designed clinical studies using sophisticated basic laboratory tools and to initiate laboratory projects to answer clinically relevant questions.

Critical Clinical Care Issues

The LSU represents a tertiary referral site for patients with severe forms of liver failure. Compulsive attention to detail is the hallmark of care in this setting; it has to start before a patient arrives at the Unit. Most patients are transferred to the LSU from other facilities. It is important to establish whether patients require airway protection, sedation and treatment of cerebral edema during transport. A line of communication needs to be established with referring physicians to ensure that all diagnostic material, including copies of films as well as histologic slides and records of diagnostic and therapeutic interventions are forwarded to the LSU. This avoids delays in diagnosis and duplication of tests. In addition, it allows more accurate assessment of disease progression.

The key clinical issue is a determination of whether a patient suffers from acute exacerbation of an underlying chronic liver disease, a subfulminant form of liver failure or fulminant hepatic failure (FHF). It is often difficult to differentiate among the three categories in late stage SALF because of incomplete history and records, because patients have already received a variety of treatments including sedatives and because they are not clinically stable to allow complete diagnostic work up. Patient record review and interviews with family members and family physician(s) are important. In general, patients are unable to directly provide any information; the majority of patients transferred to the LSU are sedated and endotracheally intubated.

Another issue presenting early following admission to the Unit is the determination of whether a patient needs a liver transplant. Standard clinical decision guidelines are used to help make that determination.[3] On the basis of the history, physical examination, initial diagnostic work-up and evaluation of the patient, an urgent decision is made as to whether the patient is a transplant candidate. For patients who are not transplant candidates, the therapeutic goal is to support them until they recover and return to their baseline clinical activity level. If patients are transplant candidates, the goal is to successfully "bridge" them to transplantation and occasionally to spontaneous recovery if their condition improves dramatically while they are waiting for a transplant.

The major determinant of severity of illness in these patients, as well as the major cause of death, is the presence of cerebral edema with brain-stem compression due to intracranial hypertension. This is primarily seen in FHF patients but can be seen, to a lesser degree, in all types of liver failure. The major problem is that intracranial hypertension often develops rapidly and without warning. Therefore, well trained staff is needed to recognize very early neurologic findings suggestive of intracranial hypertension. This requires constant observation with frequent comprehensive neurological examinations. Patients with signs of brain-stem dysfunction require placement of intracranial pressure monitors. To attempt to quantitate the degree of global neurologic dysfunction, we determine the Glasgow Coma Scale and more specifically follow the degree of brain-stem dysfunction using the Comprehensive Level of Consciousness Scale (CLOCS).[4,5] In addition, the level of encephalopathy is assessed using the Trey-Davidson classification.[6]

Patients with FHF, in general, tend to progressively deteriorate and die, or get better and fully recover spontaneously or undergo liver transplantation and recover. Subfulminant and acute-on-chronic failure patients are more likely to stay in the critical care Unit longer and develop sepsis and multiorgan failure. In this group, there is no need for constant senior physician presence around the clock at the bedside, but rather a steady long term plan of care. This includes attention to detail, nutritional and metabolic support, treatment of sepsis as well as attempts to prevent various other complications, i.e., stress ulcers, variceal bleeding, adult respiratory distress syndrome, etc.

The types of patients described above cover the extremes of SALF and require critical care. In the fulminant group, the overwhelming consideration is prevention of irreversible brain damage and, in the acute-on-chronic patients, prevention of death from sepsis and multiorgan failure. All currently used therapeutic modalities need to be standardized and management algorithms need to be developed for use of drugs and other interventions based on review of best practices from the literature. Periodic reviews and changes need to be made following updated literature data analysis. Experimental methods of treatment can only be introduced in a Unit where standard therapy is strictly controlled, allowing elimination of large numbers of variables and making data analysis meaningful.

References

1. Lee WM. Acute liver failure. N Engl J Med 1993; 329:1862-1868.
2. Demetriou AA. Support of the acutely failing liver. Medical Intelligence Unit. Austin, R.G. Landes, 1994.
3. O'Grady JG, Alexander GJ, Haylar KM et al. Early indicators of prognosis in fulminant hepatic failure. Gastroenterology 1989; 97:439-445.
4. Teasdale G, Jennett B. Assessment of coma and impaired consciousness. Lancet 1974; 2:81-82.

5. Stanczak DE, White JG, Gouview WD, Moehle KA, Daniel M, Novack T, Long CJ. Assessment of level of consciousness following severe neurological insult. J Neurosurg 1984; 60:955-960.
6. Trey C, Davidson CS. The management of fulminant hepatic failure. In: Popper H, Schaffner F, eds. Progress in Liver Disease. Volume 3. New York. Grune & Stratton, 1970.

CHAPTER 11

Other Experimental Therapeutic Strategies: Liver Gene Therapy

Jacek Rozga, Helen Lilja, Walid Arnaout, Achilles A. Demetriou

Somatic human gene therapy has been a topic of surging interest during the last several years. Initial results of experiments altering transformed cell lines to produce foreign gene products appeared promising. Early gene therapy techniques utilized naked DNA transfer and had limited long term success.[1]

In general, use of naked DNA transfection was associated with poor long-term expression. Later gene therapy experiments concentrated on the use of retroviral vectors to introduce genetic information into cell lines.[2,3] Though some investigators have questioned the efficacy of this strategy, there has been extensive and elaborate work to define systems of retroviral transduction and develop more efficient and safer ways to utilize these virus particles as a means of gene therapy.[4-7]

Retroviral delivery of genetic information to hematopoietic cells has been widely described.[3] This system has also been used for embryological mapping of cells and to determine the stage of loss of pluripotentiality. With increasing awareness of the ability of retroviruses to transduce primary cells, many investigators focused on the hepatocyte.[7-14] However, genetic alteration of these cells both in vitro and in vivo has been disappointing, chiefly because of problems with long term expression of foreign genes and difficulty with expansion of the transduced hepatocyte populations. In vivo hepatocyte transduction has been investigated by a number of groups, including our own.

In general, two major strategies have been utilized in gene therapy:

1. Cell transplantation—removal of cells from a subject with a specific genetic defect, correction of the defect, expansion of the corrected cell population and re-introduction of the cells into the same subject;
2. Direct gene delivery—delivery of a missing or corrected gene into the appropriate organ in vivo.

Transplantation of Genetically-Altered Cells

Genetically-altered hepatocyte transplantation would be the logical choice for the treatment of specific genetic defects of liver function. In general, it has been felt that currently used retroviral vectors should pose no problems to patients participating in human trials of gene therapy. However, transduction of hepatocytes with retroviral vectors has had limited success; similarly, hepatocyte transduction with adenoviral vectors, although remarkably efficient and successful, is short-lived. Renewed interest in the transplantation of genetically altered fetal hepatocytes may lead to the development of novel methods of liver gene therapy. However, thus far, this area has not been adequately explored. Similarly, development of

Support of the Acutely Failing Liver, 2nd ed., edited by Achilles A. Demetriou and Frederick D. Watanabe.

primary hepatocyte lines able to proliferate while retaining their differentiated functional characteristics may lead to cells which can be targeted for gene therapy studies.

Some investigators have utilized recombinant myocytes as a method of gene therapy.[15] While use of myocytes may offer an advantage over hepatocytes in terms of ease of harvesting and delivery of the corrected cells, long term survival and stable expression of products by these cells, in general, has been poor. Endothelial cells could potentially be used for cell transplantation. However, in spite of some early promising work in this area, no effective endothelial cell transplantation method has evolved.[16] Many investigators have turned to another cell type which appears to be more easily transduced, the fibroblast. Fibroblasts are easily harvested and cultured and have been shown to have synthetic and secretory capabilities in a constitutive manner. The limitation of using transduced fibroblasts is their inability to exhibit regulated secretory function. There have been several reports in the literature describing successful in vitro and in vivo function of retrovirally transduced fibroblasts.[17,18] Palmer et al showed in vitro and in vivo that genetically altered fibroblasts could correct factor IX deficiency.[19] We chose fibroblasts as a target cell for retroviral transduction and re-transplantation. In this section, we describe a method of retrovirally-mediated human factor IX (hF-IX) transduction of rat and human skin fibroblasts with stable expression in vitro and in vivo.[20]

Cell Donors

Syngeneic Lewis rats were purchased from Sprague-Dawley Inc., Indianapolis, IN. Wistar rats belonged to an inbred strain maintained at our animal facility. All animals were housed in a 12 hour light and 12 hour dark environment with controlled temperature and humidity. Animals were fed standard rat chow and drank tap water ad libitum. Human skin fibroblasts were harvested from circumcision tissue specimens obtained from normal infants.

Cell Preparation

Rat and human fibroblasts were obtained using a primary explant technique: Skin was harvested from the dorsum of rats after shaving and cleaning with betadine. The skin was then cut into 2-3 mm^2 sections and placed in a concentrated solution of amphotericin B (25 mg/ml), streptomycin (10 mg/ml) and penicillin G (10,000 U/ml) for 20 minutes. Sections were then placed onto 100 mm^2 plastic culture dishes and allowed to adhere in a 37°C, 5% CO_2 humidified incubator for 20 minutes. Approximately 5-8 sections were placed onto each dish. Following adherence, Dulbecco's modified Eagle's medium (DMEM) with 30% fetal bovine serum (FBS) and 1X Antibiotic-Antimycotic Solution (Sigma Chemical Co.), were gently added to the dishes. The skin primary explants were allowed to grow to 30% confluence, and at that point tissue sections were removed and the remaining fibroblasts were trypsinized (Trypsin/EDTA, Sigma Chemical Co.) and replated on Falcon T-75 tissue culture flasks.[20]

Transduction Vectors

We used an amphotropic Moloney murine leukemia virus vector. One vector contained the genes conferring resistance to hygromycin b (HB) and β-galactosidase (β-gal) production. The other vector contained the HB resistance gene ('*hph*') and the gene responsible for the production of hF-IX. The two retroviral transduction vectors were a generous gift of Immunex Corp. (Seattle, WA). The infectivity of the vectors was first assessed by transduction of NIH 3T3 cells.

Virus Harvesting

Producer cell lines containing the virus were grown to confluence on T-75 tissue culture flasks using DMEM and 10% FBS. At cellular confluence, media were changed and the following day (16-24 hours later) were collected, spun at 2000 rpm for 5 minutes and the supernatant was filtered (0.22 m) and stored at -70°C. Supernatant infectivity was tested on control NIH 3T3 cells. Highly infective batches (> 106 virions/ml) of supernatant were used and the others were discarded. Virus supernatants were thawed in a 37°C water bath.

Transduction

Fibroblast cultures were allowed to grow to 50% confluence on T-75 tissue culture flasks. Control cultures were then placed in 10 ml of DMEM/10% FBS, while cells to be transduced were incubated overnight with 10 ml of the specific vector supernatant in DMEM, 10% FBS and polybrene (hexadimethrine bromide, 4 mg/ml, Sigma Chemical Co., St. Louis. MO). Following overnight transduction, media were again changed and cells were allowed to grow to confluence.

Cell Selection

At confluence, cells were trypsinized and replated onto T-25 tissue culture flasks in DMEM, 10% FBS and HB (250 mg/ml, Calbiochem Corp., La Jolla, CA). After four days of culture in the selection media, all nontransduced fibroblasts died. Transduced cells were maintained in the selection media for the remainder of the experiment. During repeated transductions, the efficiency of transduction was approximately 0.1% for the hF-IX vector and < 0.01% for the β-Gal vector.

In Vitro Harvest

Selected resistant clones were allowed to reach confluence in T-75 flasks. DMEM was then changed to McCoy's media, 10% heat inactivated serum (HIS) and 50 ng/ml vitamin K. Following a 24 hour incubation period, the media were removed, centrifuged for five minutes at 2000 rpm to remove cells and debris and stored at -70°C until assayed. Cells were then trypsinized, pelleted and viable cells were counted using the trypan blue exclusion test. Aliquots were stored at -70°C for future use.

Factor IX Analysis

Human factor IX was assayed in both media and plasma samples using ELISA (Asserachrom IX: Antigen kit from Diagnostica, Stago, France). ELISA plates (Dynatech, Alexandria, VA) were sensitized with hF-IX-specific rabbit antibodies. Diluted rat plasma and in vitro media samples were then assayed. Ortho-phenylene-diamine (O.P.D.) was used for color development. Absorbency of each well was read at 490 nm using a Bio-Rad Micro-Plate Reader Model 3550. By standard curves of known hF-IX samples, a curve fit program provided by Bio-Rad Microplate Manager related sample absorbency to hF-IX concentration.

Retroviral infection of NIH 3T3 cells by harvested viral supernatants showed that the hF-IX/'hph' retrovirus infected approximately 0.1% of cells exposed to the virus at 50% confluence. The second vector, β-Gal/'hph', showed poor transduction efficiency (< 0.001%). Frozen (-70°C) viral supernatants remained effective in transducing both NIH 3T3 cells and primary fibroblasts. Supernatants from a single harvest kept their infectivity up to four months after freezing. Polybrene was used to enhance the efficiency of viral transduction. Transduction of primary fibroblast cultures appeared to be slightly less effective than that of NIH 3T3 cells. Transduction with the β-Gal/'hph' vector produced such low levels of transduction that it was abandoned.

Selected HB-resistant fibroblast clones and normal control fibroblasts were grown to confluence and assayed for appearance of hF-IX in the media after overnight incubation in HIS and vitamin K. While some investigators report high levels of hF-IX production in vitro, our data suggest a more modest level of factor production.[20] Control fibroblasts showed no production of hF-IX as measured by the ELISA technique. Similarly, normal human fibroblasts did not produce hF-IX, while transduced human fibroblasts produced 315 ng/10^6 cells. Factor IX activity was measured by the one-stage clotting assay. All transduced cell lines retained their ability to produce similar significant quantities of hF-IX in vitro for up to six months after transduction (mean value for all cell lines: 150.9 ± 8.2 ng/10^6 cells). Storage at $^-$70°C for two months did not have an effect on in vitro hF-IX production by the transduced fibroblasts. Polymerase chain reaction (PCR) was carried out both in vitro and in vivo. Transduced, selected clones showed incorporation of the '*hph*' proviral sequence into the fibroblast genome in vitro and in vivo. Southern blot analysis confirmed the PCR data. These data demonstrate that retroviral vector transduction results in stable cell transformation.

In the experiments described above, a retroviral vector was used to successfully transduce primary rat and human skin fibroblasts with hF-IX. Expression of hF-IX by cells not generally capable of secreting this product offers hope for possible future use of the transduced cells to correct specific genetic defects. Also encouraging was the finding of incorporation of proviral DNA sequences into the host genome, suggesting that long term production of the missing protein is possible. In addition, immunohistochemical staining demonstrated that transplanted cells were still expressing the foreign gene up to two months following transplantation. However, it was disappointing to see that although fibroblasts produced significant amounts of hF-IX in tissue culture and stained positive for hF-IX in the peritoneal cavity of transplanted recipients, there was no detectable hF-IX in the plasma of transplanted animals. This could be due to a number of possibilities which require further investigation.

Direct Gene Delivery

Recently, several investigators succeeded in delivering functional genes to hepatocytes in vitro, but a number of formidable issues remain to be resolved to achieve stable, long term expression of these genes. These include improved transduction efficiency, expansion of the population of transduced cells and development of techniques of re-transplantation in the original donor. Attempts to deliver genes directly to cells in vivo have been reported with short term success. Here, we describe a novel technique of selective in situ perfusion of the rat liver, which was successfully used to carry out retrovirus-mediated transduction of hepatocytes in vivo.

Technique

The following technique was used (Figs. 11.1 and 11.2). The abdomen was entered through a long midline incision. The ligamentous attachments of the liver were divided. We found helpful the use of an operating microscope at this early stage. Both costal margins were lifted using a U-shaped "retractor" made out of a paper clip. Next, while applying pressure downward on the liver with two fingers, the inferior caval vein was dissected between the liver and the diaphragm using curved microforceps. A 4-0 silk (Ethicon) tie was placed around the inferior caval vein and both ends passed through a 5 cm long polyethylene tubing (PE 50) and secured with a hemostat. A tourniquet was thus created which would be used to occlude the suprahepatic inferior caval vein. The intestine was exteriorized and wrapped in a gauze soaked in warm normal saline. The infrahepatic inferior caval vein was dissected above the right renal vein and a branch of the phrenic vein, where a second tourniquet was placed and left untied. The hepatic artery was dissected with great care to

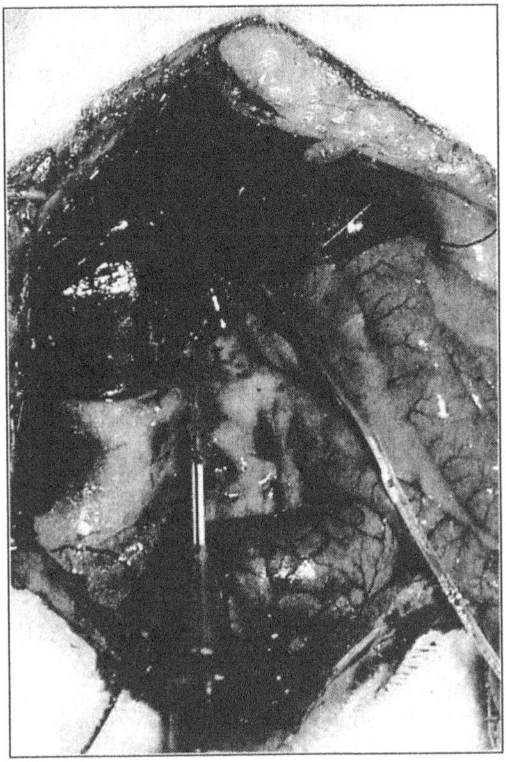

Fig. 11.1. Photograph of an actual perfusion of the posterior liver lobes.

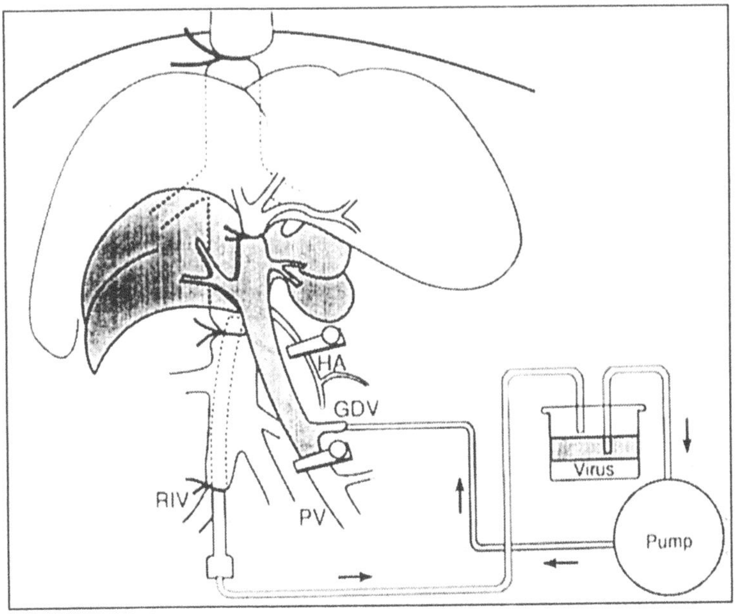

Fig. 11.2. Diagrammatic depiction of the in vivo perfusion system (RIV, right iliac vein; HA, hepatic artery; GDV, gastroduodenal vein; PV, portal vein).

avoid contraction of the vessel. Subsequently, the gastroduodenal vein, which is the most proximal tributary of the portal vein, was dissected and tied off distally using a 6-0 silk. A second 6-0 silk tie was placed at the orifice of the gastroduodenal vein. The two ligatures were then pulled, the lumen of the vein was opened and silastic tubing (Medical-Grade Tubing, Dow Croning, O.D. 0.025 in) prefilled with saline was inserted into the gastroduodenal vein and thereafter into the main trunk of the portal vein. The tubing was secured by tying the proximal ligature. It is important to note that, in spite of the large size of the tubing, it is soft and easy to manipulate. Cannulation is bloodless and can be accomplished rapidly without interrupting the portal flow. The liver hilum was exposed and the portal venous branch supplying the anterior liver lobes was dissected and ligated. This was done to stimulate DNA synthesis (regeneration) in the remaining perfused liver lobes. In controls, the portal venous branch was not ligated. Instead, a tourniquet was used to isolate the anterior liver lobes during perfusion. The right iliac vein was dissected, distally ligated and punctured with a 16 gauge catheter (Flash Cath., Travenol, Dearfield, IL). The site of puncture was secured with a sling and the tip of the catheter was advanced into the inferior caval vein above the second tourniquet, i.e., in the vicinity of the hepatic veins.

Perfusion

The vessels were occluded in the following order: hepatic artery, portal vein below the gastroduodenal vein, portal venous branch supplying the two anterior liver lobes, infrahepatic inferior caval vein and suprahepatic inferior caval vein after injection of 3-5 ml of lactated Ringer solution to wash the posterior liver lobes out of blood. Recirculating perfusion with viral supernatant (1×10^6 virions/ml, 40 ml) or vehicle alone (controls) was carried out at the rate of 3 ml/min using a roller pump (Masterflex, Barnant Co, Barrington, IL), as shown in Figure 11.2, for 15 minutes. Subsequently, the tourniquets and microclips were removed, inflow and outflow tubings were withdrawn and the gastroduodenal and right iliac veins were ligated. Normal blood flow to the posterior liver lobes was re-established.

Experimental Design

Sixteen Sprague-Dawley rats underwent in situ perfusion of the posterior liver lobes. They were divided into four groups: Group I rats (n = 4) underwent portal branch ligation and perfusion with 'hph' virus at the same time. Group II rats (n = 4) underwent portal branch ligation and perfusion with 'hph' virus 20 hours later. Group III rats (n = 4) were perfused with 'hph', without portal branch ligation. Group IV rats (n = 4) underwent portal branch ligation followed, 20 hours later, by perfusion with vehicle alone. One rat from each group was killed at 24 hours, with anesthetic overdose, for histological evaluation of the perfused liver lobes. All remaining rats were evaluated seven days after perfusion by isolating hepatocytes from the perfused liver lobes, culturing them and selecting with HB.

Selection with Hygromycin B

Cells were plated on 35 mm collagen-coated plates (1×10^5 cells/plate). During the attachment period, Dulbecco's modified Eagle's medium without glutamine but with 10% fetal bovine serum was used. After three hours, the medium was changed to hormonally-defined serum-free RPMI 1640, which was then replaced every 24 hours. Cultures were maintained at 37°C in a humidified atmosphere (5% CO_2 and 95% air). After three days in culture, cells were selected with HB (LD99: 500 mg/ml) and their viability was assessed after four days, following fixation in methanol and Giemsa staining. Percent viability and transduction were calculated.

Vascular isolation of the liver in situ without systemic and splanchnic venous decompression was routinely carried out in approximately 15-16 minutes. At 24 hours following liver perfusion, there was no histologic evidence of hepatocyte injury by light microscopic morphologic criteria. In groups I, II and IV, the nonligated, previously perfused liver lobes appeared normal histologically and numerous cells in mitosis were observed. After transduction with the 'hph' virus, 9.2 ± 0.5% of cultured hepatocytes in Group I survived HB treatment. In controls perfused with vehicle alone (Group IV), no resistant hepatocytes to HB selection were seen (0.6 ± 0.2% viable cells). Similar results were obtained in nonligated Group III rats perfused with the 'hph' virus (2.8 ± 0.9% viable cells; $p>0.05$ compared with Group IV controls). In contrast, significant transduction was achieved in Group II, in which the nonligated regenerating liver lobes were perfused with the virus at 20 hours, the expected peak of DNA synthesis. In the livers of these animals, 16.0 ± 1.0% hepatocytes were resistant to HB (statistical significance was $p < 0.003$ when compared to the other groups).

The isolated perfused liver has been used to study hepatic metabolism and as a model for regional cancer chemotherapy. We have now utilized selective liver lobe perfusion in vivo to carry out targeted delivery of foreign genes to hepatocytes. Several animal models of in vivo liver perfusion have been described. Most of them, however, are technically demanding and require oxygenation and/or shunting to prevent splanchnic venous stasis. In contrast, our model is simple, rapid, safe and, with some practice, it is associated with neither perioperative mortality nor damage to the liver. The main reason for the simplicity of this technique is that the perfusion period is kept short (not longer than 20 minutes). Apart from technical details, another major difference with other methods is the ability in the described model to carry out selective isolation and perfusion of the posterior liver lobes followed by reperfusion and establishment of normal blood flow. Use of portal branch ligation to induce DNA synthesis was found to be less invasive and equally effective as partial hepatectomy.

Hepatocyte cultures prepared from rats infused with an amphotropic retrovirus carrying the '*hph*' gene demonstrated transduction of a significant number of hepatocytes. It is important to note that livers perfused with the vector at the time of expected peak DNA synthesis (20 hours), yielded hepatocyte cultures with significantly higher resistance to hygromycin b toxicity when compared to those transduced at the time portal branch ligation was carried out. In the absence of portal branch ligation, almost no resistance to hygromycin b was seen (< 1%). We therefore demonstrated that anterior liver lobe ligation is an effective technique of enhancing perfused liver cell transduction, presumably by stimulating DNA synthesis. This is in agreement with our experience with retroviral transduction in partially hepatectomized rats in vivo, as well as with the results of in vitro studies by Miller et al suggesting that active DNA synthesis is needed for efficient uptake, integration and expression of retroviral DNA to targeted cells.[4,16]

Use of selective in situ perfusion rather than direct intraportal infusion of viral particles was carried out because, in pilot experiments using this virus, we were unable to obtain hepatocyte infection following direct infusion of virus particles into the portal vein. This is probably due to inadequate uptake of virus particles by the hepatocytes following a single passage through the liver. Once through the liver, the virus can be taken up by other tissues. It would be theoretically possible, however, to achieve hepatocyte infection by direct infusion of virus into the portal vein by infusing a very large number of virions in a constant infusion over a long period of time. Infusion of such large numbers of virus particles in the whole body may be detrimental. We believe that for future clinical therapy, it would be advantageous to avoid the need for partial hepatectomy to induce enhanced hepatocyte DNA synthesis. Selective (lobar) occlusion, either by ligation of portal venous branches or insertion of balloon-tipped catheters, can be achieved in humans (or large animals) to induce enhanced DNA synthesis. In addition, growth factor infusion through catheters in portal vein branches could be carried out alone or in combination with occlusion. We therefore believe that the technique of portal vein ligation as a means of stimulating hepatocyte DNA synthesis has greater clinical relevance than partial hepatectomy. In conclusion, we developed a novel method of selective liver lobe perfusion combined with portal vein branch ligation which results in successful retrovirus-mediated transduction of adult rat hepatocytes in vivo. Isolated organ perfusion could be adapted for use in large animals or humans and, therefore, it may be a potentially useful tool for treating specific inborn errors of hepatic metabolism.

Gene Transfer in Fetal Rat Hepatocytes

In contrast to adult hepatocytes, fetal hepatocytes are thought to be highly proliferative, a feature which should facilitate gene transfer requiring active DNA synthesis. However, data on fetal hepatocyte proliferative activity are limited and very few reports exist on retrovirally-mediated gene transfer in rat FH.[21,22]

Fetal cell suspension was obtained using a modification of mechanical separation and nonperfusion collagenase/DNAse digestion originally described by Devirgiliis et al.[23] Fetal and adult rat hepatocytes were plated onto 60 mm Permanox plates (NUNC) coated with rat tail collagen type I (Becton Dickinson Collab. Biomed Products) in 3 ml of a hormonally and chemically defined medium. It consisted of a 1:1 (vol/vol) mixture of Ham F12 and Williams E medium supplemented as recommended by Isom and Georgoff, except that dexamethasone concentration was reduced to 10^{-7} mol/l.[24] Insulin and proline were present at 0.6 mg/l and 32.25 mg/l, respectively. During the first 4 hours of culture, 5% FBS was added to the culture medium to encourage cell attachment. Serum-free medium was used every 24 hours. Cells were cultured at 37°C in a humidified air/5% CO_2. The rate of DNA-SA (synthetic activity) was determined by measuring incorporation of ^{3}H-thymidine into the trichloroacetic acid (TCA)-precipitable fraction of the cell lysate. Human recombinant HGF was purchased from R&D Systems (Minneapolis, MN). A working solution was prepared as recommended by the supplier. The culture conditions (cell density, HGF dose, timing of maximal response) were determined in preliminary studies (data not shown). Accordingly:

1. Fetal hepatocytes and adult hepatocytes were seeded at a cell density of 20 x 10^3/cm^2 and 10 x 10^3/cm^2, respectively.
2. HGF (20 ng/ml) was added at 4 hours after plating and then renewed every 24 hours as the medium was changed.
3. At 4 hours before the time of DNA assay, 10 ml of ^{3}H-thymidine (83.2 Ci/mmol; DuPont NEN Products, Boston, MA) were added per 3 ml of culture medium per dish. Cells cultured in the absence of HGF served as controls.
4. DNA-SA was determined in fetal hepatocytes and adult hepatocyte cultures at 52 and 72 hours postplating, respectively.

In some experiments, 15 mmol/l hydroxyurea was added to distinguish between DNA repair and incorporation of ^{3}H-thymidine caused by DNA repair. At the time of assay, cells were washed 3X with PBS and scraped in the same buffer. After centrifugation, cell pellets were incubated in 0.5 mol/L NaOH (1 ml per 1.6 x 10^6 cells) for 30 minutes at 37°C while being vortexed every 10 minutes. A 150 ml aliquot of cell lysate was mixed with 500 ml distilled water and 350 ml 25% TCA and incubated at 37°C for 10 minutes. After centrifugation at 10,000 x g, the precipitated material was washed three times with 10% TCA. Radiolabeled DNA (TCA precipitate) was hydrolyzed by 10% TCA (30 minutes at 90°C). After centrifugation at 10,000 x g, the supernatant was collected and mixed with 4 ml of scintillator (Ready Safe Beckman, Fullerton, CA) and radioactivity was measured in a liquid scintillation counter. DNA-SA was expressed in cpm/min/mg of total protein lysate. Protein concentration was determined using a Pierce BCA Protein Assay Reagent Kit (Pierce Chemicals, Rockford, IL).

An amphotropic retrovirus containing the β-galactosidase gene expressed under 5'LTR internal promoter was used. It was made by transfection of 293 T cells and kindly provided as frozen stock by Dr. T Engstrand, UCLA. Fetal and adult hepatocytes (20 days of gestation) were plated as described above. Four hours after plating, the medium was removed and replaced by 1 ml of medium containing the retroviruses. Polybrene was added at a concentration of 8 mg/ml. After 4 hours, the cells were placed in serum-free medium, which was renewed after 24 hours. At 48 hours posttransfection, the cells were stained with X-gal as described by Sambrook et al.[25] Percent transduction was calculated.

Viability of fetal hepatocytes after isolation was always greater than 95%, as determined by trypan blue exclusion. Cell yield was approximately 1 x 10^7/g liver. On FACS analysis, almost 90% of cells expressed a-fetoprotein and albumin and therefore were considered to be hepatocytes and their committed progenitors. Fetal hepatocytes had a high base proliferative activity. In fact, in nonstimulated fetal hepatocytes, it was significantly (ten-fold) higher than the maximal response induced by HGF in adult hepatocyte cultures (Fig. 11.3). After retroviral transduction (n = 6), 5-15% of nonstimulated fetal hepatocytes expressed β-galactosidase (Fig. 11.4). In contrast, virtually none of the cultured adult hepatocytes stained positive for X-gal.

Thus, we demonstrated that nonstimulated fetal hepatocytes exhibit a high level of spontaneous DNA synthetic activity. It was found to be ten times greater than the peak response induced by HGF in adult hepatocyte cultures. This explains, at least partially, why in subsequent studies fetal hepatocytes were found to be susceptible to retrovirally-mediated gene transfer. In this regard, our data seem to be in conflict with the findings of Richardson et al, who reported that fetal liver cells were virtually immune to amphotropic retroviral transduction.[21] In those studies, however, cells were procured from embryonic livers which were harvested at day 15 of gestation, a time when more than 95% of fetal liver cells are known to be hematopoietic progenitors.[26] Although Hatzoglou et al were also unable to infect the fetal rat liver with amphotropic retroviruses, it is worth noticing that

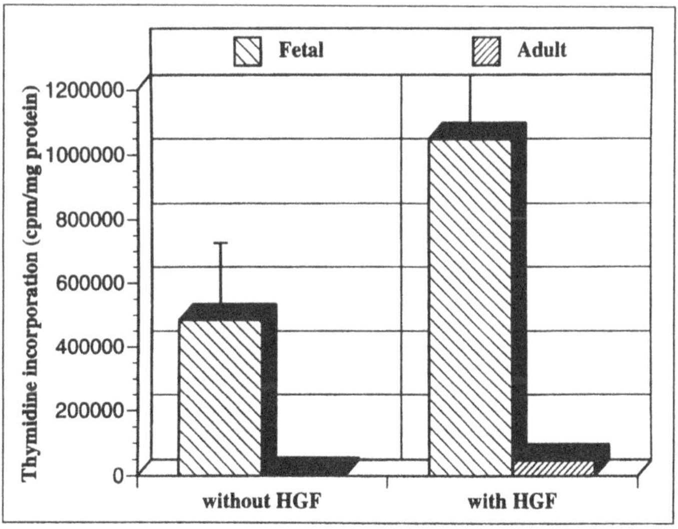

Fig. 11.3. Comparison of DNA synthetic activity in quiescent and HGF-stimulated cultured fetal and adult rat hepatocytes. Data are shown as means ± SD. The observed differences were statistically significant at $p < 0.005$ or less.

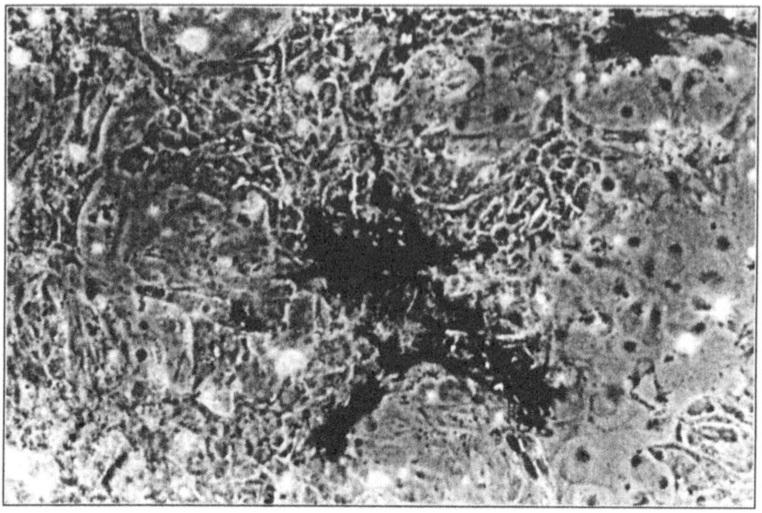

Fig. 11.4. Microphotograph depicting cultured fetal rat hepatocytes which were exposed to an amphotropic retroviral vector carrying the *lac-Z* gene encoding for *E. coli* β-galactosidase. Successfully transduced cells stained positive for X-gal (dark blue).

they injected the viral suspension intraperitoneally and, therefore, viral particles could reach the liver only after uptake by the systemic circulation.[22]

In summary, we have confirmed earlier observations that fetal rat liver from the late gestation period is highly enriched with committed hepatocyte progenitors. These cells exhibit high basal DNA synthetic activity, which makes them a good target for gene transfer studies requiring active DNA synthesis.

References

1. Cline MJ, Stang H, Mercola K et al. Gene transfer in intact animals. Nature 1980; 284:422-425.
2. Price J, Turner D, Cepko C. Lineage analysis in the vertebrate nervous system by retrovirus-mediated gene transfer. Proc Natl Acad Sci USA 1987; 84:156-160.
3. Cepko C. Lineage analysis and immortalization of neural cells via retrovirus vectors. In: Boulton AA, Baker GB, Campagnoni AT, eds. Neuromethods, vol 16, Cliffton, New Jersey: Humana Press, 1989:177-218.
4. Miller A, Buttimore C. Redesign of retrovirus packaging cell lines to avoid recombination leading to helper virus production. Mol Cell Biol 1986; 6:2895-2902.
5. Overell RW, Weisser KE, Cosman D. Stably transmitted triple-promoter retrovirus-mediated gene transfer. Proc Natl Acad Sci USA 1987; 84:156-160.
6. Bender MA, Palmer TD, Gelinas RE et al. Evidence that the packaging signal of Moloney murine leukemia virus extends into the gag region. J Virol 1987; 61:1639-1646.
7. Moscioni AD, Rozga J, Neuzil DF et al. In vivo regional delivery of retrovirally mediated foreign genes to rat liver cells: Need for partial hepatectomy for successful foreign gene expression. Surgery 1993; 113:304-311.
8. Ledley FD, Darlington GJ, Hahn T et al. Retroviral gene transfer into primary hepatocytes: Implications for genetic therapy of liver-specific functions. Proc Natl Acad Sci USA 1987; 84:5335-5339.
9. Wolff JA, Yee J-K, Skelly HF et al. Expression of retrovirally transduced genes in primary cultures of adult rat hepatocytes. Proc Natl Acad Sci USA 1987; 84:3344-48.
10. Wilson JM, Jefferson DM, Roy Chowdhury J et al. Retrovirus-mediated transduction of adult hepatocytes. Proc Natl Acad Sci USA 1988; 85:3014-3018.

11. Wilson JM, Johnston DE, Jefferson DM et al. Correction of the genetic defect in hepatocytes from the Watanabe heritable hyperlipidemic rabbit. Proc Natl Acad Sci USA 1988; 85:4421-4425.
12. Anderson KD, Thompson JA, DiPietro JM et al. Gene expression in implanted rat hepatocytes following retroviral-mediated gene transfer. Somat Cell Mol Genet 1989; 15:215-227.
13. Armentano D, Thompson AR, Darlington G et al. Expression of human factor IX in rabbit hepatocytes by retrovirus-mediated gene transfer: Potential for gene therapy of hemophilia B. Proc Natl Acad Sci USA 1990; 87:6141-6145.
14. Rozga J, Moscioni AD, Neuzil D et al. A modal for directed foreign gene delivery to rat liver cells in vivo. J of Surg Res 1993; 52:209-213.
15. Wolff JA, Malone RW, Williams P et al. Direct gene transfer into mouse muscle in vivo. Science 1990; 247:1465-1468.
16. Yao SN, Wilson JM, Nobel EG et al. Expression of human factor IX in rat capillary endothelial cells: Toward somatic gene therapy for hemophilia B. Proc Natl Acad Sci USA 1991; 88:8101-8105.
17. Scharfmann T, Axelrod JH, Verma IM. Long-term in vivo expression of retrovirus-mediated gene transfer in mouse fibroblasts into mice: Implications for gene therapy. Science 1987; 236:714-718.
18. St Louis D, Verma IM. An alternative approach to somatic cell gene therapy. Proc Natl Acad Sci USA 1988; 85:3150-3154.
19. Palmer TD, Thompson AR, Miller AD. Production of human factor IX in animals by genetically modified skin fibroblasts: Potential therapy for hemophilia B. Blood 1989; 73:438-445.
20. Neuzil D, Holzman M, Rozga J et al. Fibroblast transplantation in rats: Transduction and function of foreign genes. J of Surg Res 1993; 54:631-637.
21. Richardson C, Ward M, Podda S, Bank A. Mouse fetal liver cells lack functional amphotropic retroviral receptors. Blood 1994; 84:433-444.
22. Hatzoglou M, Moorman A, Lamers W. Persistent expression of genes transferred in the fetal rat liver via retroviruses. Somatic Cell and Molecular Genetics 1995; 4:265-273.
23. Devirgiliis L, Dini L, Di Pierro A et al. An improved nonperfusion method for the isolation and purification of rat fetal and neonatal hepatocytes. Cell & Molecular Biology 1981; 27:687-693.
24. Isom HC, Georgoff I. Quantitative assay for albumin-producing liver cells after simian virus 40 transformation of rat hepatocytes maintained in chemically defined medium. Proc Natl Acad Sci USA 1984; 8:6378-6384.
25. Sambrook J, Fritsch EF, Maniatis T. Molecular Cloning. A Laboratory Manual. Cold Spring Harbor N.Y, 1989:1666.
26. Fukumoto T. Possible developmental interactions of hematopoietic cells and hepatocytes in fetal rat liver. Biomed Res 1992; 13:385-392.

CHAPTER 12

Other Experimental Therapeutic Strategies: Hepatocyte Transplantation

Helene Lilja, Nikolaos Arkadopoulos, Susumu Eguchi, Raoul J. Rosenthal, David A. Moscioni, Achilles A. Demetriou, Jacek Rozga

Transplantation of hepatocytes, injection of various hepatocyte extracts and hepatocyte culture supernatants, have been shown to prolong survival in animals with D,(+)-galactosamine-induced liver injury and animals with acute liver ischemia.[1-3] Hepatocyte transplantation into the spleen, fat pads, dorsal fascia, lungs, pancreas and portal vein has been carried out resulting in partial correction of specific liver metabolic defects, i.e., hyperbilirubinemia due to lack of UDP-glucuronyl transferase for bilirubin (Gunn rat) and/or morphologic demonstration of differentiated hepatocytes at various ectopic sites.[4-7] Several studies have attempted to provide a matrix and solid supports, including biodegradable scaffolds, for transplanted hepatocytes to facilitate cell engraftment and enhance neovascularization.

The hepatocyte transplantation literature is extensive but often contradictory and incomplete. In general, there are several problems with many hepatocyte transplantation studies:

1. Use of experimental animal models which are neither standardized nor validated and thus are difficult to reproduce (i.e., galactosamine-induced liver failure, temporary liver ischemia).
2. Use of indirect assays of transplanted hepatocyte function (i.e., measurement of serum bilirubin rather than appearance of bilirubin conjugates in the bile of transplanted Gunn rats).
3. Reliance on morphologic evidence of presence of cells in the animal at ectopic sites without demonstration of function (i.e., correction of functional metabolic defects).
4. Lack of long-term success; most reported studies demonstrate only a transient effect (up to 3-4 weeks) of transplanted hepatocytes even in congeneic strains of animals.
5. Reports of potentially significant preliminary observations without follow-up studies.

For a hepatocyte transplantation method to be useful, the technique must be simple, result in early engraftment and allow transplantation of an adequate number of hepatocytes, and transplanted hepatocytes should express differentiated functions in vivo. We have developed methods of hepatocyte transplantation addressing and solving some of the problems outlined above and we propose a series of studies to attempt to resolve the remaining issues. Most of the work outlined here has dealt with the correction of specific genetic defects. This is an essential step toward development of methods of treating acute liver failure in animals. It provides a better experimental animal model for assessing the function of transplanted hepatocytes and quantitating their effects, and allows better assessment of engraft-

Support of the Acutely Failing Liver, 2nd ed., edited by Achilles A. Demetriou and Frederick D. Watanabe.

ment and long term survival and differentiated function of these cells. Once these techniques are developed, they could be used to treat acute liver failure. Another potential limitation of hepatocyte transplantation is that the number of hepatocytes needed to provide metabolic support is greater than that needed for correction of a specific genetic defect. In addition, it probably takes a period of time for transplanted cells to engraft and function; therefore, the utility of this approach in the urgent treatment of acute liver failure remains to be established. Our own bias is that acute liver failure will probably be best treated with extracorporeal liver support systems, and hepatocyte transplantation may be more applicable to the treatment of genetic defects of liver function. However, in spite of all its limitations, hepatocyte transplantation remains a powerful experimental, and possibly therapeutic, tool which may have a role in the management of acute liver failure.

For the past fourteen years, our laboratory has been working towards developing techniques of experimental isolated hepatocyte transplantation. Our aim has been to collect data serving as a useful information base for efforts to develop methods of clinical hepatocyte transplantation. Specifically, we described techniques of intraperitoneal, intraportal and intrasplenic cell transplantation and developed reliable experimental animal models to test transplanted cell function.[7-18] Additionally, we developed methods of immunomodulation of cells prior to transplantation, resulting in prolongation of their in vivo survival, methods of studying transplanted cell proliferation, immunohistochemical techniques for examining transplanted cell morphology and methods of harvesting hepatocytes from large animals.[7,8,11,13,19-21] These studies are summarized below.

Correction of Specific Genetic Defects

Tissue culture studies demonstrated that hepatocyte survival and differentiation in vitro was enhanced by utilizing a collagen matrix. We applied similar principles in developing in vivo hepatocyte transplantation techniques.[6,7] We hypothesized that attachment of isolated hepatocytes to a collagen matrix prior to transplantation will improve transplanted cell survival and expression of differentiated hepatocyte function. Normal rat hepatocytes were harvested by collagenase portal vein perfusion and were attached to type I collagen-coated dextran microcarriers.[7] Microcarrier-attached hepatocytes are shown in Figure 12.1 following 3 hours of attachment. Microcarrier-attached normal rat hepatocytes were transplanted into animals with specific genetic defects of liver function: rats unable to conjugate bilirubin (Gunn) or synthesize albumin (Nagase analbuminemic rats; NAR). We were able to demonstrate survival and function of transplanted normal hepatocytes by:

1. The appearance of bilirubin conjugates (Fig. 12.2), the increase in total bilirubin in the bile (Fig. 12.3) and a decrease in serum bilirubin levels in transplanted Gunn rats (Fig. 12.4).
2. An increase in plasma albumin levels in cyclosporin-immunosuppressed allogeneic NAR recipients (Fig. 12.5).

Treatment of Acute Liver Insufficiency

Microcarrier-attached normal rat hepatocytes were transplanted into rats with acute liver insufficiency following 90% partial hepatectomy. A dramatic improvement in recipient survival and prevention of postoperative hypoglycemia in rats transplanted four days prior to 90% partial hepatectomy were noted. There were no significant differences in animal survival between rats given allogeneic vs. those receiving syngeneic hepatocytes. It takes several days for intraperitoneally transplanted allogeneic hepatocytes to be rejected; we thus concluded that only early hepatocyte survival is important in prolonging animal survival in these animals, by providing metabolic support during a time of acute liver insufficiency

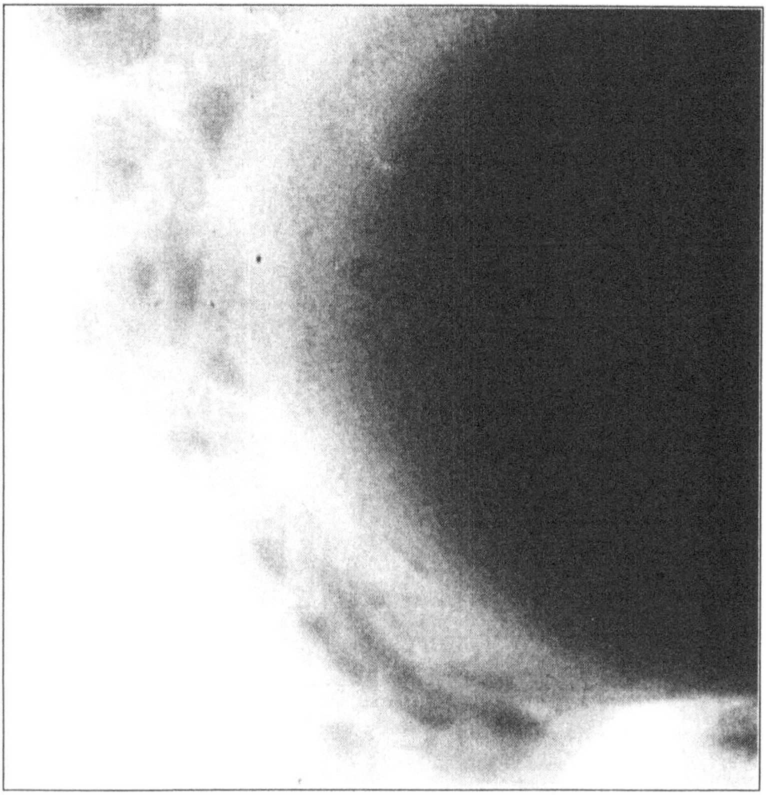

Fig. 12.1. Microcarrier-attached hepatocytes following three hours of attachment.

until the native liver recovers. These data demonstrate survival and function of transplanted hepatocytes in vivo resulting in metabolic support of animals with acute liver insufficiency.

Other Considerations

Transplanted Hepatocyte Proliferation

In experiments in which animals undergoing 90% partial hepatectomy demonstrated improved survival following hepatocyte transplantation, it was noted that transplantation of only a small number of hepatocytes (1-2% of the total liver mass), resulted in improvement in animal survival.[8] One possible explanation is that transplanted hepatocytes proliferated at their ectopic sites. Studies were undertaken to determine whether transplanted microcarrier-attached hepatocytes proliferate in the peritoneal cavity of syngeneic rat recipients. Microcarrier-attached normal hepatocytes were transplanted intraperitoneally into syngeneic Lewis rats. Three days later, transplanted rats underwent 70% partial hepatectomy and 24 hours later they were injected with ^{3}H-thymidine intravenously. One hour later, all rats were killed with ether and their livers and peritoneal cavity aggregates were excised and assayed for total DNA and ^{3}H-thymidine incorporation into DNA. There was a significant increase in ^{3}H-thymidine incorporation into DNA in the regenerating liver remnants when compared to sham-operated animal livers (13,150 ± 1,080 vs. 860 ± 119 dpm/mg DNA; $p < 0.01$); similarly, there was a significant increase in ^{3}H-thymidine uptake

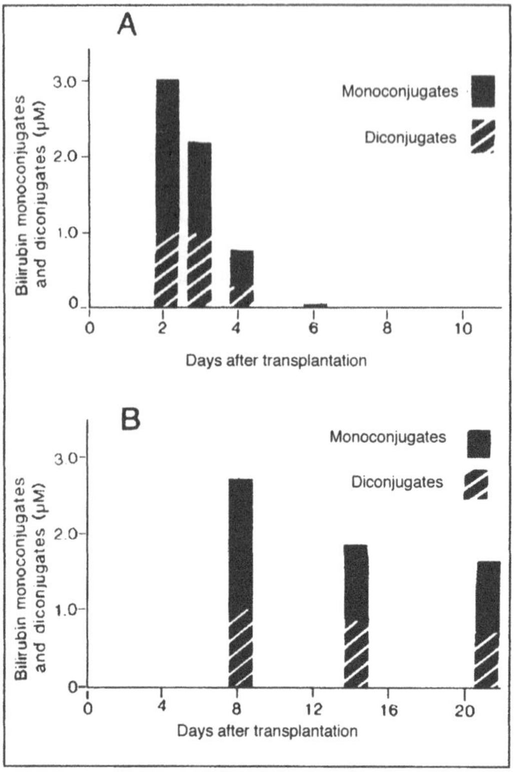

Fig. 12.2. Appearance of bilirubin conjugates.

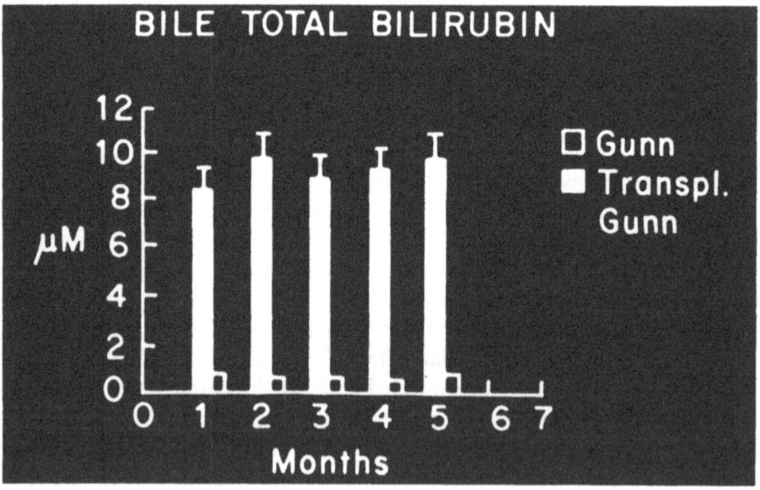

Fig. 12.3. Increase in total bilirubin in the bile.

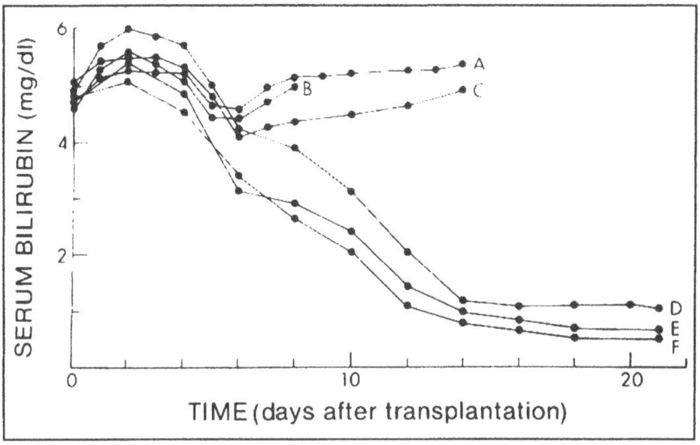

Fig. 12.4. Decrease in serum bilirubin levels in transplanted Gunn rats. (A-C): Transplanted Gunn rats; (E-F): Nontransplanted controls.

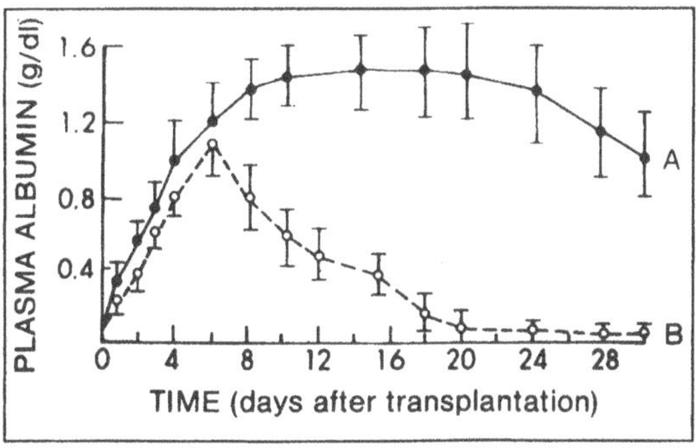

Fig. 12.5. Increase in plasma albumin levels in cyclosporin-immunosuppressed allogeneic NAR recipients. (A): Transplanted NAR rats; (B): Nontransplanted controls.

by peritoneal aggregates in partially hepatectomized rats when compared to uptake in sham-operated rats (4,500 ± 610 vs. 729 ± 165 dpm/mg DNA; $p < 0.02$). Data from this and other experiments carried out in our laboratory, suggested that transplanted liver cells proliferate in vivo. Experiments were then carried out to determine which cells in the peritoneal cavity aggregates proliferate. Dividing cells were labeled using either ^{3}H-thymidine autoradiography or staining for bromodeoxyuridine. Sections were also stained with a specific hepatocyte probe (anti-albumin antibody) in an attempt to identify proliferating hepatocytes. Using such a "double label" technique, we were able to identify a large number of proliferating cells in the aggregates; however, only a rare hepatocyte was shown to proliferate. Our data suggest that although cells within the peritoneal cavity aggregates proliferate in response to partial hepatectomy, most proliferating cells were not differentiated hepatocytes; we were thus unable to demonstrate hepatocyte proliferation to a significant extent using immunohistochemical/nuclear labeling techniques.

Morphologic Studies

Intraperitoneally transplanted microcarrier-attached hepatocytes in rats form well-vascularized aggregates.[7,8] The morphology of the aggregates was examined using light and electron microscopy and immunohistochemical techniques. Using standard indirect immunoperoxidase methods, formalin-fixed, paraffin embedded sections of NAR rat peritoneal cavity aggregates were stained with anti-rat albumin antibody. A large number of hepatocytes were identified staining for albumin up to 28 days following transplantation in cyclosporin-treated NAR rats (Fig. 12.6). Most hepatocytes were found in clusters embedded in the connective tissue matrix among microcarriers, indicating that cell-cell contact may be important for transplanted cell survival. Use of the NAR rat recipient allows better definition of the albumin-stained hepatocytes due to low background staining. Electron microscopic examination of the aggregates revealed presence of giant nucleoid-containing peroxisomes which are characteristic for normal rat hepatocytes and formation of microvilli and canaliculi, suggesting a high level of morphologic differentiation. Scanning electron microscopy revealed clusters of cells and microcarriers and newly formed collagen on the surface of the microcarriers (Fig. 12.7).

More recently, we performed a series of experiments with intraperitoneal transplantation of microcarrier-attached hepatocytes in pigs. In this species, morphological appearance of peritoneal aggregates was even more "liver-like" than in rats (Fig. 12.8).

Use of Biodegradable Collagen Matrix

To develop a clinical method of intraperitoneal hepatocyte transplantation, it would be advantageous to utilize a totally biodegradable matrix as a vehicle for cell transplantation to eliminate foreign body reaction in the recipients. We thus attached normal isolated hepatocytes to thin (2 mm) slices of type I collagen sponges and transplanted 1×10^7 cells into NAR rats by inserting several sponge slices (total: 140 mg) into the peritoneal cavity through a small abdominal incision. This resulted in a significant increase in serum albumin levels. Rats with collagen sponge slices alone had no changes in serum albumin (0.05 ± 0.01 mg/dl);

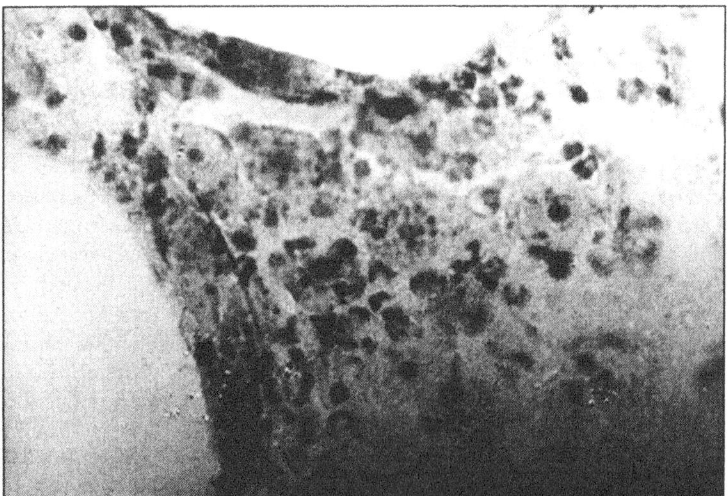

Fig. 12.6. Hepatocytes stained for albumin up to 28 days following transplantation in cyclosporin-treated NAR rats.

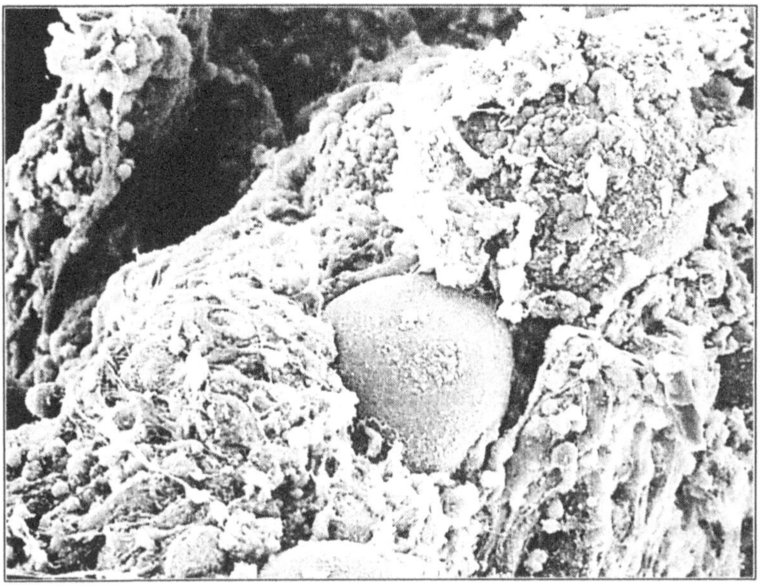

Fig. 12.7. Clusters of cells and microcarriers and newly formed collagen on the surface of the microcarriers, by scanning electron microscopy.

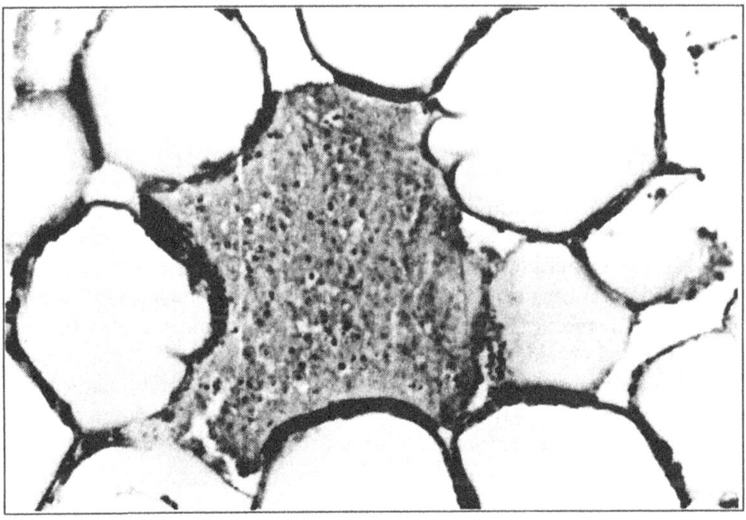

Fig. 12.8. Clusters of hepatocyte aggregates recovered from the peritoneal cavity of the pig which received intraperitoneal injection of microcarrier-attached porcine hepatocytes two weeks earlier. The recipient was maintained on daily cyclosporin A (25 mg/kg i.m.).

16 rats (50%) transplanted with hepatocyte-sponge slices, demonstrated a significant linear (peak level day 6; 2.5 ± 1.1 mg/dl, $p < 0.01$) increase in serum albumin. Western blot analysis confirmed the SDS-PAGE data. However, this method does not appear to be as effective as the microcarrier technique. Most cells in the center of the sponges were not viable and only a relatively small number of viable cells was found on the surface of the sponge. Studies were also carried out using totally degradable type I collagen microcarriers which did not appear to be as effective as the collagen sponge matrix.

Modulation of Transplanted Cell Antigenicity

Ultraviolet Irradiation

Exposure to ultraviolet (UV) irradiation has been shown to prolong pancreatic islet allograft survival. The purpose of this study was to examine the effect of UV irradiation on the survival and function of normal hepatocytes in vivo. Microcarrier-attached normal rat hepatocytes were pretreated with UV irradiation (600 J/m^2) prior to transplantation into allogeneic NAR recipients. Transplanted hepatocyte survival and function were assessed by serial measurements of plasma albumin levels. There was a significantly prolonged elevation of plasma albumin levels in rats transplanted with UV-treated cells when compared to rats transplanted with sham-irradiated cells. Our results demonstrate that pretreatment with UV irradiation prolongs survival and function of allogeneic intraperitoneally-transplanted, microcarrier-attached, adult normal rat hepatocytes.

Encapsulation

Cell encapsulation has been shown to prolong transplanted cell function because the semipermeable membrane reportedly immunoisolates the cells. Transplantation of encapsulated cells has been carried out by Sun et al, employing 25,000 MW alginate/poly-L-lysine (ALP).[22] We carried out a series of experiments using ALP-encapsulated microcarrier-attached hepatocytes.[10] We hypothesized that combining microcarrier and encapsulation techniques would allow cell attachment to a matrix and, additionally, immunoisolate the cells. In vitro experiments were first carried out to examine transport of macromolecules across the ALP capsule. Encapsulated microcarrier-attached hepatocytes were cultured under standard conditions in defined serum-free media. Albumin release into the media was measured by an enzyme-linked immunoabsorbance assay (ELISA). At 24 hours, nonencapsulated, microcarrier-attached hepatocytes released more albumin than encapsulated ones into the medium (25.0 ± 2.0 vs. 4.1 ± 1.2 ng/ml; $p < 0.02$). Subsequently, in a series of in vivo experiments, NAR rats were transplanted either with microcarrier-attached allogeneic hepatocytes or with encapsulated microcarrier-attached hepatocytes, both groups without cyclosporin A immunosuppression. No statistically significant differences were noted. A very strong inflammatory response was seen in the peritoneal cavity of animals transplanted with encapsulated microcarrier-attached hepatocytes. No such reaction was seen in controls either injected with encapsulated microcarriers alone or animals transplanted with syngeneic microcarrier-attached encapsulated hepatocytes. Encapsulated allogeneic hepatocytes induced a significant pericapsular inflammatory reaction, probably due to release of cytokines and other inflammatory response mediators.

Human Hepatocyte Studies

Our laboratory has carried out experiments utilizing human hepatocytes in a series of studies resulting in the development of methods for harvesting, testing and cryopreserving human hepatocytes for subsequent clinical use. Human hepatocyte function was tested by transplanting these cells into the peritoneal cavity of athymic (nude) rats with specific inherited defects of liver function. In a way, these animals served as in vivo "test tubes" for evaluating human hepatocyte function.[23] Rat recipients were either hyperbilirubinemic Gunn rats or NAR analbuminemic rats. Rats were made genetically immunodeficient by interbreeding with athymic rats with inherited T cell deficiency. There was no morphologic evidence of rejection in the immunoincompetent recipients, whereas immunocompetent rats demonstrated rejection within five days of transplantation. Nude Gunn rat recipients demonstrated excretion of bilirubin glucuronides in bile for up to thirty days and reduction in serum bilirubin levels. In recipient NAR rats, plasma albumin levels increased from a pretransplant level of 0.025-0.05 mg/ml to 3.9-4.8 mg/ml and remained at nearly that level for thirty days. These experimental studies thus demonstrate that human hepatocytes can perform well differentiated normal liver functions in vivo.

Optimal Technique

Although significant progress has been achieved in isolated hepatocyte transplantation, the optimal site of implantation has not been determined. Many investigators have transplanted hepatocytes at sites away from the liver, including spleen, kidney, abdominal cavity, thymus, testis, brain, pancreas, lungs, subcutaneous tissue, muscle, fat pads and other locations.[1-8,24-26] Although the spleen and peritoneal cavity have been most frequently used, there are reasons to believe that transplanted hepatocytes, similarly to auxiliary liver (auto) grafts, undergo atrophy when implanted ectopically outside the portal stream. In addition, the intrasplenic route is limited by the small number of cells transplanted. In 1986, Vroemen et al observed that, during intrasplenic injection, the cell suspension was flowing via the splenic vein into the portal system. [6] Similar findings were reported by Gupta et al.[25] We therefore examined direct intraportal hepatocyte transplantation. We were hoping to obtain better cell engraftment, survival and function, because:

1. Transplanting hepatocytes into the unique liver architecture allows interaction with other hepatocytes and nonparenchymal cells;
2. Proximity to hepatocyte-specific growth and differentiation factors creates an environment particularly conducive to hepatocyte engraftment and function;
3. Locally released mitogens and portal-borne hepatotrophic factors can increase transplanted cell numbers;
4. The liver may be an immunologically privileged site;
5. In the liver, hepatocytes are able to secrete bile directly into the biliary tree.

In the past, intraportal infusion of isolated hepatocytes produced severe liver damage, due to the occlusion of the portal blood supply by transplanted cells.[4] In addition, early mortality resulted from cell aggregates passing to the cardiopulmonary circulation, portal vein thrombosis and portal hypertension. We have overcome these complications by developing a method of intraportal injection of isolated hepatocytes in a single cell suspension selectively into specific liver lobes, but not others, which allows portal decompression following transplantation.[9] Use of this technique has resulted in improved cell engraftment, absence of thrombosis in the portal venules and minimal to no injury to the liver. We have subsequently transplanted hepatocytes repeatedly by means of an indwelling catheter system connected to the portal venous branch and designed a method of stimulating transplanted hepatocytes to proliferate by means of portal blood occlusion of the nontransplanted liver lobes. [12,19] The latter strategy resulted in near-complete correction of analbuminemia in NAR rats[13] and in profound and sustained lowering of blood cholesterol levels in Watanabe hyperlipidemic (WHHL) rabbits.[14]

Selective Intraportal Hepatocyte Transplantation

NAR rats underwent midline laparotomy under anesthesia.[9] The portal venous branch supplying the anterior or posterior liver lobes was isolated and a loose tourniquet was placed around it. The gastroduodenal vein was identified and cannulated using silastic tubing. A vascular clamp was temporarily placed distally on the portal vein. The tourniquet around the portal venous branch was secured and 2×10^7 cells (5 ml) were infused into the nonoccluded liver lobes over a 1 minute period. Vascular flow to the liver was re-established by removing the clamps and tourniquet at the end of the infusion.

NAR rats were transplanted via the gastroduodenal vein into either the posterior (Group I) liver lobes (30% of the total liver mass; n = 10) or the anterior (Group II) liver lobes (70% of the total liver mass; n = 5). Freshly isolated single cell suspensions of normal Sprague-Dawley hepatocytes were infused over two minutes (2×10^7 cells). Controls (Group III) consisted of NAR recipients transplanted with allogeneic NAR donor hepatocytes (n = 8). All animals were immunosuppressed with cyclosporin A in weekly doses of 10 mg/kg i.m. for the duration of the experiment. Blood samples were obtained from the rat tail vein at days 0, 4, 7, 14, 21 and 28 for serum albumin determinations. Two animals from each group were killed at 7, 14 and 21 days each. All remaining animals were similarly killed at 28 days. Tissue specimens from both the posterior and anterior liver lobes and from lungs were immunostained utilizing rabbit anti-rat albumin IgG antibodies and an immunoperoxidase avidin-biotin peroxidase complex (Vector Laboratories, Burlingame, CA) method. Serum albumin levels were determined by SDS/10% polyacrylamide gel electrophoresis (SDS-PAGE); purified rat albumin and normal Sprague-Dawley rat serum were used for reference. The specificity of the assay was confirmed by carrying out immunotransblots using rabbit anti-rat albumin antiserum. Quantitative analysis of albumin in serum samples was carried out by sandwich ELISA, using a modified double antibody assay.

A significant ($p < 0.003$) and sustained increase in plasma albumin concentration occurred in all NAR rats transplanted with normal Sprague-Dawley rat hepatocytes (Table 12.1, Fig. 12.9). At 28 days posttransplantation, Group I had greater than a 5-fold increase in plasma albumin levels. Group II showed greater than 12-fold increase in plasma albumin levels. Group III demonstrated no significant change in plasma albumin levels at 28 days posttransplantation. The ELISA findings were confirmed by the 10% SDS-PAGE. The protein band increase seen on the SDS-PAGE was shown to be albumin by protein immunoblot. No significant changes in total serum protein were found following transplantation with either normal or NAR hepatocytes in all groups of recipients.

Liver sections stained by the immunoperoxidase method were examined for the presence of albumin-stained cells. Sections of nontransplanted NAR rats demonstrated a small

Table 12.1. Total serum albumin concentration in NAR rats transplanted with normal Sprague-Dawley rat hepatocytes

	Baseline	Day 4	Day 7	Day 14	Day 21	Day 28
Group I (n = 10)	4.46 ± 0.45	6.10 ± 1.0	8.14 ± 1.4*	14.91 ± 3.0*	19.75 ± 4.1*	24.68 ± 8.2*
Group II (n = 5)	2.70 ± 0.2	6.27 ± 1.0*	13.05 ± 5.4*	22.84 ± 6.8*	45.54 ± 14.4*	34.93 ± 12.0*
Group III (n = 8)	3.53 ± 0.1	3.53 ± 0.1	3.80 ± 0.1	3.62 ± 0.1	3.52 ± 0.1	3.53 ± 0.2

Data are expressed as mg/dl (mean ± SEM); * $p < 0.01$ (Students t-test).

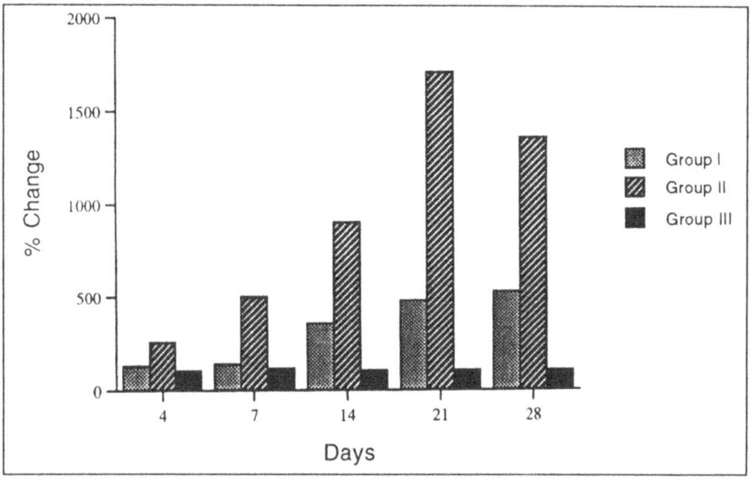

Fig. 12.9. Percent change in plasma albumin concentrations from baseline for groups I, II, and III as determined by sandwich ELISA.

number of albumin-positive cells, occurring mostly as singlets or doublets. Donor Sprague-Dawley rat liver sections stained virtually homogeneously for the presence of albumin within hepatocytes. Experimental animals transplanted with normal hepatocytes had increased numbers of albumin-positive cells in their livers when compared to nontransplanted NAR rats.

The method utilized in this series of experiments was unique in that only selected lobes of the liver were infused, rather than the entire liver parenchyma. It is felt that this critical variation in the technique of intraportal transplantation allows portal decompression through the nontransplanted lobes, thus decreasing the mortality observed in previously published intraportal transplantation studies and development of portal hypertension. It is also important to note that this technique was not associated with cellular embolization in the lungs, as has been previously reported following direct intraportal transplantation.

In conclusion, we have developed a simple technique for intraportal hepatocyte transplantation in rats, which has resulted in successful intrahepatic cell engraftment and function.

Repeated Intraportal Hepatocyte Transplantation

Since only a limited number of hepatocytes can be injected directly into the portal vein without causing liver injury, we developed a method of repeated infusion of hepatocytes into the portal system.[12] Normal Sprague-Dawley rat hepatocytes were harvested by portal vein EDTA/collagenase perfusion. Recipient analbuminemic (NAR) rats were divided into two groups. Group I (n = 7) underwent subcutaneous placement of a Port-a-Cath reservoir, connected to silicone tubing tunneled subcutaneously into the abdomen and placed in the gastroduodenal vein at the portal vein junction; rats were infused with 3 ml of normal saline on days 1, 3, 5, 8, 10. Group II (n = 7) underwent placement of a reservoir as above, and received repeated infusions of 5×10^6 normal rat hepatocytes in 3 ml of saline on the same days. All rats were given cyclosporin A (10 mg/kg BW, i.m. three times a week). Plasma albumin levels were determined by ELISA. Animals were killed on day 28, their livers were fixed, sectioned and stained for albumin as in the earlier studies. Plasma albumin levels (mg/dl; mean ± SEM) are shown in Table 12.2. There was a significant ($p < 0.01$) increase in plasma albumin levels in transplanted rats with no change in controls. Serum AST and ALT

Table 12.2. Plasma albumin levels in analbuminemic (NAR) rats transplanted with single (2×10^7) and multiple ($5 \times 10^6 \times 6$ doses) injections of isolated normal albumin producing rat hepatocytes

	Baseline	Day 7	Day 14	Day 21	Day 28
Controls	3.33 ± 0.35	3.46 ± 0.40	3.50 ± 0.43	3.13 ± 0.25	3.90 ± 0.4
Single injection	4.83 ± 0.64	8.38 ± 1.08	16.43 ± 3.00	19.75 ± 4.10	24.70 ± 7.1*
Multiple injections	3.33 ± 0.20	5.80 ± 0.76	14.00 ± 2.70	28.84 ± 4.60	47.14 ± 7.5**

Data presented as mg/dl (mean ± SEM). Statistical significance (Students t-test): * $p < 0.05$; ** $p < 0.01$.

levels remained normal in both groups throughout the study. Immunohistochemical staining demonstrated presence of large numbers of albumin-producing hepatocytes only in the livers of Group II rats; histologically there was no liver damage in either group.

In summary, this method offers the option of repeated cell infusions if the number of transplanted hepatocytes is insufficient to obtain a desired functional effect following a single administration, and if transplanted hepatocytes fail to survive or are rejected. Repeated administration of cells via an implanted device may have a place in both ex vivo liver-directed gene therapy and in transplantation of allogeneic isolated hepatocytes.

Hepatocyte Transplantation Under a Regeneration Response

All methods of hepatocyte transplantation described to date have not demonstrated either sustained long term liver support or normalization of a genetic liver defect. This has been primarily due to inadequate quantities of hepatocytes transplanted. Even in the above described studies in NAR rats, the number of transplanted cells was small (6 batches of 5-6 million cells represented less than 3% of the host liver mass). We therefore combined selective intraportal hepatocyte transplantation with a liver regeneration stimulus in an attempt to increase the number of functioning transplanted cells.[13]

We previously demonstrated that, in the rat, by performing portal branch ligation in which the two anterior lobes (70% of total liver mass) were permanently occluded from the circulation, a regeneration response resulted in the remaining nonoccluded lobes similar to that seen after 70% hepatectomy.[19] In contrast to Hamaguchi et al,[27] who transplanted hepatocytes shortly after partial hepatectomy, we first allowed hepatocytes transplanted in the posterior liver lobes to engraft and then, by ligating the portal blood inflow to the nontransplanted anterior liver lobes, induced a regeneration response in the transplant-bearing part of the liver.[13]

Adult male NAR rats underwent midline laparotomy. The liver hilum was exposed, the portal venous branch supplying the two anterior lobes dissected and a silk tie placed around it. A 25 gauge needle was inserted into the portal vein via the gastroduodenal tributary and, while restricting flow to the anterior liver lobes, a single cell suspension of normal hepatocytes (2×10^7 cells/3 ml of culture medium) was infused into the right and omental lobes (Group I). Following cell infusion, the gastroduodenal vein was ligated, the portal venous ligature was released (but left in place) and animals were allowed to recover. Two weeks after transplantation, the rats' abdomen was re-opened and the portal venous ligature was applied. Control groups included:

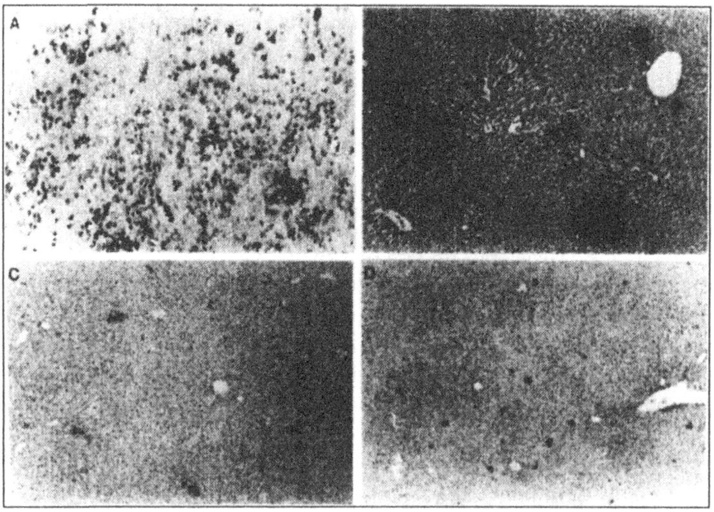

Fig. 12.10. Photomicrographs of liver sections obtained from Group I (A), Group II (B), Group III (C), and Group IV (D) experimental NAR rats 86 days following transplantation, immunohistochemically probed for albumin-positive hepatocytes.

Group II: Hepatocyte transplantation using normal hepatocytes plus sham-portal blood ligation;

Group III: Hepatocyte transplantation using NAR hepatocytes plus portal blood ligation; and

Group IV: Sham-hepatocyte transplantation (i.e., saline infusion) plus portal branch ligation.

The experiment was terminated 86 days posttransplantation.

Serum albumin levels in rat Groups III and IV showed no significant changes throughout the experimental period (Group III: 0.005 ± 0.001 g/dl, day 0 vs. 0.003 ± 0.001 g/dl, day 86; Group IV: 0.004 ± 0.001 g/dl, day 0 vs. 0.004 ± 0.001 g/dl, day 86 after transplantation). Group II transplanted NAR rats showed a continuous rise in serum albumin concentration throughout the experimental period following hepatocyte transplantation; by 86 days posttransplantation it was 0.260 ± 0.072 g/dl, an approximate 90-fold increase compared to Groups III and IV (p=0.001). In all Group I rats, serum albumin levels increased dramatically (300-600 fold) following portal branch ligation (1.78 ± 0.20 g/dl vs. 0.260 ± 0.072 g/dl in Group II rats, p=0.001). Histological examination of tissue sections demonstrated a remarkable expansion in the number of albumin-positive cells in Group I animals, when compared to the remaining three groups. Quantitation of albumin-positive cells between Group I and II animals showed a significant 8-fold increase in positive hepatocytes in animals which subsequently underwent a regeneration stimulus (PBL): Group I: 22.6 ± 7.5% vs. Group II: 2.75 ± 2.7% ($p < 0.005$). Control Groups III and IV showed less than 1% albumin-positive hepatocytes in representative tissue sections (Fig. 12.10).

In this study, we obtained for the first time a near-total correction of a genetic defect in liver function. At 86 days posttransplantation, five out of eight Group I NAR rats had serum albumin levels between 2.0 g/dl and 2.5 g/dl (in normal rat liver cell donors it was 3.5 ± 0.4 g/dl). There was a marked increase in the population of albumin-positive cells in the livers of transplanted rats undergoing liver regeneration (approximately 23% of the cell population). However, the mass of donor liver cells transplanted into the posterior lobes of recipient rats represented not more than 9% of the total population of hepatocytes within these lobes. This has been demonstrated in parallel experiments, where carboxyfluorescein-labeled cells were transplanted in the posterior liver lobes and then identified by fluorescence microscopy in tissue sections and by FACS analysis following collagenase digestion of the liver tissue.[11] The greater than expected increase in cell numbers observed in NAR animals may represent preferential selection for growth of normal hepatocytes. This suggestion is supported by an earlier study, in which we demonstrated that NAR rats undergoing 70% partial hepatectomy had significantly diminished DNA synthetic activity when compared to normal Sprague-Dawley rats undergoing an identical procedure.[28] It is also possible that donor hepatocytes develop micro-emboli within the transplanted lobes, allowing preferential nutrition, oxygenation and exposure to portal-borne hepatotrophic factors. In addition, some of the host hepatocytes deprived of these factors might undergo atrophy or even apoptosis. Another interesting observation is that in all transplanted NAR rats studied, albumin levels continued to increase in the serum long after the regeneration stimulus had ceased and the livers had reached their original mass. Whether this is related to the above proposed preferential survival and/or proliferation of normal albumin-producing hepatocytes, or to continuing overexpression of the albumin gene in the transplanted cells, remains to be investigated.

In summary, we have described a method of regional transplantation of a small number of normal hepatocytes into the posterior lobes of NAR rats and subsequent induction of regeneration of the transplanted lobes by portal venous branch ligation. Combined use of these two techniques resulted in pronounced expansion of the population of normal hepatocytes and significant elevations in serum albumin levels, which approached near normal values in several animals by the end of the experiment (86 days).

Hepatocyte Transplantation in Watanabe Hyperlipidemic Rabbits

The success of the above studies in Nagase analbuminemic rats encouraged us to use the same strategy in Watanabe heritable hyperlipidemic (WHHL) rabbits.[14] The genetic defect in these animals is a counterpart of familial hypercholesterolemia in humans. Homozygous animals have a mutation in the low density lipoprotein (LDL) receptor gene and express almost no functional LDL receptors. As a result, intermediate density lipoproteins and LDL accumulate in the plasma, leading to accelerated atherosclerosis and premature death.

Under intravenous ketamine/xylazine anesthesia, the abdomen was entered through a midline incision. The portal venous branch supplying the right lateral liver lobe was dissected and punctured using an 24 gauge Venocath (Becton Dickinson Vascular Access, Sandy, UT). A 2-0 silk was tied around the catheter to ensure infusion of cells selectively into the right lateral lobe (20% liver mass). New Zealand White rabbit hepatocytes (2×10^8) were infused over a 5 min period. The catheter and the silk tie were subsequently removed and local hemostasis was achieved. Before closing the abdomen, the portal venous branch supplying the anterior liver lobe was dissected and a silk thread was placed around it and left untied. One week following cell transplantation, all animals were re-operated and PBL was completed. Control WHHL rabbits were transplanted with hepatocytes and one week later were subjected to sham-PBL (Group II) or received an injection of heparinized saline and one week later were subjected to PBL (Group III). All rabbits were maintained on daily cyclosporin A (25 mg/kg p.o.) for the duration of the experiment. The experiment was terminated at 150 days posttransplantation. In order to visualize LDL receptor-positive hepatocytes, all recipient livers were perfused with fluorescein (DiI)-labeled LDL. Tissue samples were then taken from both transplant-bearing and nontransplanted liver lobes and immediately frozen in OCT with liquid nitrogen for subsequent fluorescence microscopic examination. Rou-

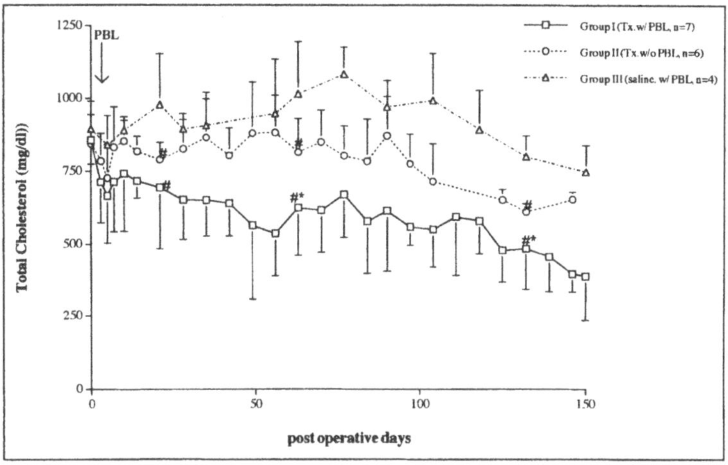

Fig. 12.11. Serum cholesterol levels in the three groups of WHHL rabbits. Group I and II rabbits were transplanted (Tx.) with allogeneic hepatocytes. Group III controls received intraportal saline injection only. Portal branch ligation (PBL) was performed on day 7 posttransplantation in Group I and III animals. Baseline serum cholesterol levels were as follows: Group I, 855 ± 81 mg/dl; Group II, 843 ± 102 mg/dl; Group III, 893 ± 97 mg/dl. Values are given as mean ± SD. * $p < 0.05$ vs. Group II, # $p < 0.05$ vs. Group III.

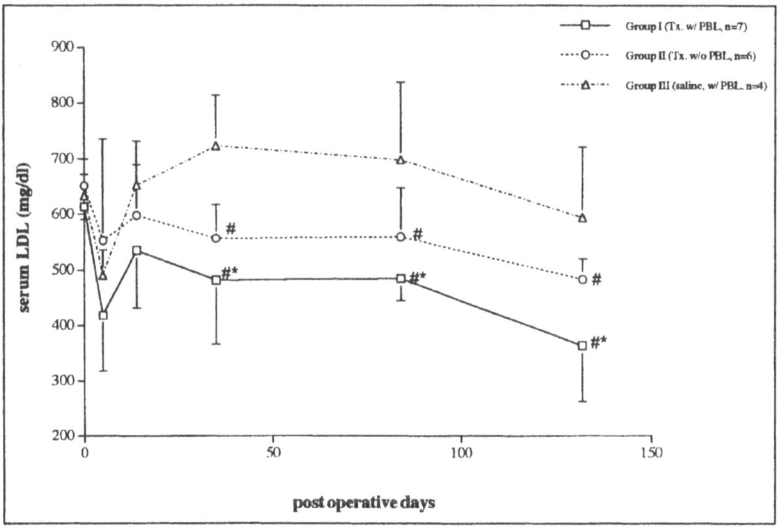

Fig. 12.12. Serum low density lipoprotein (LDL) levels in the three groups of WHHL rabbits. Group I and II rabbits were transplanted (Tx.) with allogeneic hepatocytes. Group III controls received intraportal saline injection. Portal branch ligation (PBL) was performed on day 7 posttransplantation in Group I and III animals. Baseline serum LDL levels were as follows: Group I, 614 ± 23 mg/dl; Group II, 651 ± 21 mg/dl; Group III, 633 ± 67 mg/dl. Values are given as mean ± SD. * $p < 0.05$ vs. Group II, # $p < 0.05$ vs. Group III.

tine staining using hematoxylin-eosin, oil-red O and histochemical demonstration of lipids were performed as well. In addition, sections from the coronary artery and aorta were examined for the presence of atherosclerosis.

In all WHHL rabbits, pretransplant blood cholesterol levels were over 800 mg/dl. A marked and sustained reduction of serum cholesterol and LDL levels was recorded in WHHL rabbits, in which hepatocyte transplantation was followed by PBL. The functional effect seen in these animals was far more pronounced than that seen in Group II rabbits (Figs. 12.11, 12.12). The livers previously perfused with DiI-labeled LDL contained more LDL-positive hepatocytes in Group I animals than in their Group II partners. Of great importance was the finding of a nearly total lack of atherosclerotic changes in the aortas of Group I WHHL animals. In randomly chosen serial sections, the total area occupied by atherosclerotic plaques was 0.092 ± 0.063 cm^2 in Group I rabbits; in control Groups II and III it was 0.480 ± 0.339 cm^2 and 0.682 ± 0.165 cm^2 respectively (quantitative analysis was performed using OPTIMAS 5.0 software; Group I vs. Groups II and III at the $p < 0.05$ level).

This study suggests that transplantation of normal allogeneic hepatocytes should be considered as a low risk therapy in familial hypercholesterolemia. It appears (at least in experimental animals) to be more effective than intrahepatic transplantation of retroviral corrected autologous hepatocytes. Our preliminary data also suggest that allogeneic hepatocyte transplantation, if applied early in life, may prevent development of atherosclerosis and its complications.

Hepatocyte Transplantation in Rats with Fulminant Hepatic Failure

We have developed a model of fulminant hepatic failure (FHF) in rats where resection of the anterior liver lobes (68% liver) is combined with ligation of the right lobe (24% liver) pedicle.[29] In this setting, rat survival depends upon the ability of the residual omental lobes (8% liver) to function and regenerate. It was found that with hydration as the only supportive measure, more than 90% of these FHF rats die within 48 hours postoperatively from severe liver failure, with no signs of liver regeneration.[12] Lack of a hyperplastic response in the remnant livers of these animals, was associated with delayed expression of tissue c-Met mRNA expression and marked elevation of plasma TGF-β1 levels. We carried out an experiment to determine whether a limited number of isolated hepatocytes transplanted ectopically (spleen) would prolong animal survival and improve and trigger regeneration in the native liver.

Inbred male Lewis rats underwent intrasplenic injection of 2×10^7 syngeneic hepatocytes; two days later, to allow engraftment, FHF was induced.[16] Control FHF rats received intrasplenic injection of saline. Survival time was determined in 10 test animals and 10 controls. Another 20 test animals and 20 controls were killed in batches of 5 each at 24, 36, 48 and 72 hours. At sacrifice, blood was collected for laboratory tests and the liver remnant was weighed and processed for light microscopy, HGF c-Met receptor mRNA expression (RT-PCR) and determination of proliferation indices (proliferation cell nuclear antigen, PCNA; bromo-deoxyuridine, BrdU).

Rats undergoing intrasplenic injections of hepatocytes lived longer than nontransplanted controls (73 ± 22 vs. 33 ± 9 hours; $p < 0.05$), had significantly greater liver remnant weights (203 ± 80 vs. $53 \pm 36\%$; $p < 0.05$) and demonstrated increased signs of cell proliferation (numerous cells stained positive for BrdU and PCNA and a mitotic index of $3.0 \pm 0.5\%$ at 72 hours postinduction of FHF). Transplanted rats had lower plasma TGF-β1 levels, improved HGF clearance and accelerated expression of HGF c-Met receptors in regenerating liver remnants when compared to nontransplanted FHF controls. In addition, transplanted rats had statistically significantly lower blood NH_3, lactate, LDH and total bilirubin levels compared to controls. At the time of death, the viable liver tissue in these rats corresponded to nearly 30% of their original liver mass. It is interesting, however, that this mass was unable to ensure the animals' permanent survival.

An important finding in this study was that cells corresponding to as little as 2% of the liver mass, transplanted ectopically, stimulated initiation and maintenance of the regeneration response in the native livers of recipient animals. It would therefore appear that transplanted hepatocytes could provide support for a patient with acute hepatic failure, not only by providing liver-specific functions, but also by enhancing the regenerative capacity of the native liver.

Transplantation of Fetal Rat Hepatocytes

In contrast to adult hepatocytes, fetal hepatocytes (FH) are thought to be less immunogenic and relatively more resistant to cryopreservation and ischemic injuries. These qualities could enhance transplanted cell engraftment. Nonetheless, only a few attempts have been made to transplant fetal liver tissue/cells in rats with liver failure and heritable disorders of liver metabolism.[30-32] Although some studies provided evidence of transplanted cell function, several issues regarding transplantation of fetal hepatocytes have not been addressed. For example, there appears to be confusion as to whether cell isolates enriched with committed hepatocyte progenitors can be obtained without resorting to use of complex, low yield methods such as immunological panning and fluorescence-activated cell sorting. In addition, intraportal injection of fetal hepatocytes has not been critically examined, even though the liver may provide a potentially immunologically privileged site and a unique liver-specific environment which could enhance transplanted cell engraftment, differentiation and expression of liver-specific functions. Finally, data on fetal hepatocyte proliferative activity are limited and very few reports exist on retrovirally-mediated gene transfer in rat fetal hepatocytes.[33]

Fetal livers were harvested from Sprague-Dawley rat donors at day 20 of gestation.[17] A cell suspension was obtained using mechanical separation and a nonperfusion, collagenase/DNAse digestion method. Most cells (87 ± 2%) exhibited morphological (electron microscopy) and molecular (expression of a-fetoprotein, albumin; FACS analysis) features of hepatocyte lineage. Male Nagase analbuminemic rats treated with cyclosporin A were transplanted with fetal hepatocytes (2×10^7 cells/rat) isolated from normal S-D rat donors. In Group I rats, fetal hepatocytes were infused selectively into the anterior liver lobes. In Group II rats, fetal hepatocytes were infused into the posterior lobes, in which regenerative response was induced after 24 hours by ligation of the portal branch (PBL) supplying the anterior lobes. Group III rats had 2×10^7 adult hepatocytes infused into the anterior liver lobes. Controls consisted of Group IV and V rats receiving intraportal infusion of saline with and without PBL. Only recipients of adult hepatocytes (Group III) and fetal hepatocytes under a regenerative response (Group II) showed elevation in serum albumin levels (ELISA) (Fig. 12.13). In these rats, immunohistochemistry confirmed the presence of albumin-positive hepatocytes.

In conclusion, we have confirmed earlier observations that fetal rat liver from the late gestation period is highly enriched with committed hepatocyte progenitors. We also showed that in adult analbuminemic rats, intrahepatically seeded fetal hepatocytes produced significant amounts of albumin, but only when a hepatic regenerative stimulus was applied during the early posttransplant period.

Hepatocyte Transplantation in Anhepatic Rats

Patients with FHF and liver necrosis often die before a liver becomes available for transplantation. Staged total hepatectomy has been carried out to remove diseased necrotic liver and stabilize patients until transplantation.[34,35] We questioned whether

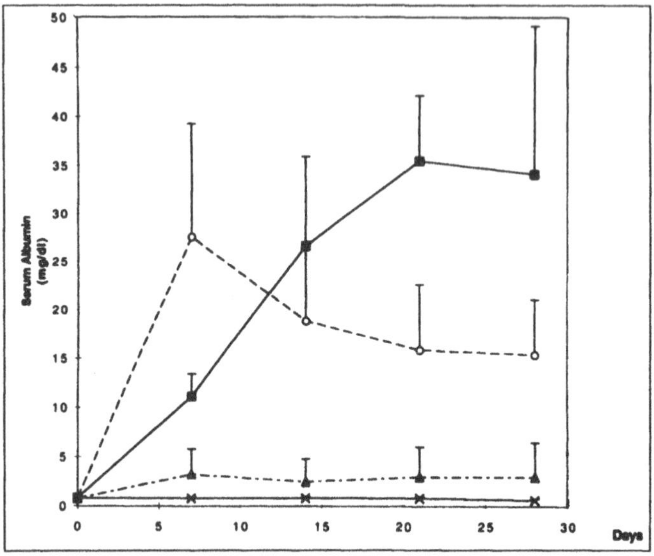

Fig. 12.13. Serum albumin levels in the experimental NAR rats at 0, 7, 14, 21 and 28 days posttransplantation. Control NAR rats were given injection of normal saline. Solid square, adult donor; o, fetal donor + PBL; ▲, fetal donor; X, control.

hepatocyte transplantation could be used in this setting to provide temporary metabolic support. To test our hypothesis, we developed a single-stage technique for total hepatectomy in rats and used it to study the effects of intrasplenic hepatocyte transplantation.[18]

Adult male Sprague-Dawley rats (150-350 g) were used. Donor hepatocytes were isolated using an in situ, two-step EDTA/collagenase digestion and enrichment through a Percoll gradient. Group I rats (n = 16) underwent intrasplenic injection of 2.5×10^7 hepatocytes. Group II rats (n = 12) underwent intrasplenic injection of saline. In all animals, 5-6 out of 6-7 splenic venous branches were permanently ligated to prevent immediate migration of transplanted cells to the liver. Both groups were treated daily with cyclosporin A (10 mg/kg, i.m.). After three days to allow cell engraftment, all rats were rendered anhepatic. Briefly, an end to side portocaval shunt (PCS) was created and the hepatic artery and bile duct were transected. Next, a 3-4 cm long piece of a 14 gauge Angiocath was introduced (by direct puncture) into the lumen of the inferior vena cava (IVC) at a level between the left renal and the right iliolumbar veins. The stent was advanced into the IVC and secured above the PCS and the liver dome. The liver was then removed, including tissue surrounding the intrahepatic portion of the IVC. Postoperatively, all rats were maintained on continuous intravenous (jugular vein) glucose supplementation (20 mg/100 g/h). Eight rats from each group were monitored to determine survival time. The remaining rats were euthanized at 12 hours post-hepatectomy for measurement of blood ammonia (NH_3) levels, prothrombin time (PT), international normalized ratio (INR) and plasma hepatocyte growth factor (HGF) and transforming growth factor β1 (TGF-β1) levels (ELISA). Spleen sections were immunostained for expression of proliferation nuclear cell antigen (PCNA). Data are expressed as means ± standard deviation. Student's t-test was used for statistical analysis.

Group I transplanted rats survived significantly longer than Group II sham-transplanted controls (34.1 ± 8.5 vs. 15.5 ± 4.8 hours, $p < 0.05$). In addition, Group I rats progressed to stage 4 encephalopathy (no righting reflex to pain stimuli) later than Group II rats (29.5 ± 7.7

vs. 10.6 ± 3.9 hours, $p < 0.05$). At 12 hours post-hepatectomy, transplanted rats had lower NH_3 levels (1535 ± 344 vs. 2137 ± 427 mg/dl, $p < 0.05$), lower PT (17 ± 1 vs. 24 ± 4 sec, $p < 0.05$) and INR (1.9 ± 0.1 vs. 3.8 ± 1.3, $p < 0.05$) than controls. Blood levels of HGF were similar in both groups (9.0 ± 2.2 ng/ml vs. 8.3 ± 4.0 ng/ml), whereas blood TGF-β1 levels were significantly lower in transplanted rats compared to controls (25.5 ± 16.0 vs. 60.1 ± 13.8 ng/ml, $p < 0.05$). Sections of transplant-bearing spleens contained numerous clusters of hepatocytes. In 50% of these animals, intrasplenic hepatocytes showed signs of proliferation (PCNA labeling index: 8-12%).

We have demonstrated that in anhepatic rats, intrasplenic transplantation of a relatively small number of allogeneic hepatocytes delayed the onset of encephalopathy and prolonged survival. Additionally, intrasplenically transplanted hepatocytes exhibited detoxifying and synthetic activity, as demonstrated by lower NH_3 levels and improved blood coagulation. We have previously shown that, in rats, the anhepatic state was associated with the progressive rise in blood HGF and TGF-β1 levels. In this study, hepatocyte transplantation had no immediate effect on blood HGF levels. It did, however, slow the rise in blood TGF-β1 levels. This growth factor profile may have been responsible for the observed transplanted cell proliferation.

In conclusion, this study suggests that hepatocyte transplantation may prove to be a useful tool for providing temporary metabolic support in anhepatic patients.

Intrasplenic Hepatocyte Transplantation in Pigs

It has been reported that after intrasplenic hepatocyte transplantation, many cells migrate to the liver without causing portal vein thrombosis.[6,25] We carried out a series of large animal experiments to examine the feasibility of the intrasplenic route and to develop a method of large animal hepatocyte transplantation using a laparoscopic technique.[15]

Young farm pigs (10-12 kg) were used as cell donors and adult pigs (35-50 kg) were used as hepatocyte recipients. Hepatocytes were harvested by a two-step EDTA/collagenase liver perfusion. Hepatocyte enrichment was achieved using a three-compartment chamber with meshes of decreased porosity and a blood cell processor (COBE 2991, Lakewood, CO) as described earlier.[21] Pelleted hepatocytes were resuspended in physiologic saline and immediately transplanted.

The following techniques of intrasplenic cell delivery were tried:

1. Retrograde infusion of cells into the splenic vein;
2. Injection of cells into the splenic artery;
3. Intrasplenic injection of cells directly into the spleen parenchyma (direct infusion along the spleen long axis during open abdominal surgery; percutaneous injection under laparoscopic control; multiple subcapsular injections of cells under laparoscopic control).

The number of cells injected varied from 2×10^9 to 10×10^9 cells. Although in all techniques tested, care was taken to avoid early leakage of cells into the portal circulation and the spleen was decompressed prior to cell injection, only direct intrasplenic injection of 2×10^9 cells was compatible with survival. However, even with such a relatively small number of cells (2% original liver mass), there was a significant risk of spleen infarction, perisplenic adhesion formation and portal vein thrombosis. The laparoscopic approach was found to be reliable, simple and safe. In general, implantation sites demonstrated chronic inflammatory changes, fibroblasts and foreign body type giant cells, with varying degrees of calcification. In some sections within a vascular channel, a sheet of loosely arranged epithelioid and spindle cells, weakly reactive with PAS, was identified. These cells also showed faint reactivity to anti-a1-antitrypsin and anti-a-fetoprotein. No cells resembling hepatocytes were found in the lungs.

In summary, the results of this experiment indicate that, even though the spleen has traditionally been considered a safe optimal site for hepatocyte transplantation, it cannot easily accommodate large numbers of cells; increasing the number of cells delivered into the spleen is associated with serious complications, including death. Further work is needed to overcome technical obstacles and new methods will have to be developed for intrasplenic cell delivery (i.e., repeated cell transplantation, injection of cells by way of multiple spleen micropunctures, intrasplenic transplantation of cells using a laparoscopically implanted Port-a-Cath and others).

Other Cell Transplantation

Enterocyte Transplantation

Columnar epithelial cells of the small intestine, commonly referred to as enterocytes, consist primarily of mature absorptive villous cells and their precursors, rapidly proliferating goblet cells. Many metabolic characteristics of isolated enterocytes have been described and their capacity for drug and substrate metabolism has been extensively studied in vitro. Hartman and Bissell demonstrated the presence of bilirubin uridine-diphosphogucuronosyl-transferase (UDPGT) in isolated rat enterocytes.[36] A series of experiments were carried out to determine whether transplantation of normal rat enterocytes into UDPGT-deficient Gunn rats can correct the underlying biochemical defect.[37]

Enterocytes were isolated from adult normal Wistar rat small intestine by using chelation-elution with a 27 mM sodium citrate buffer to dissociate the cells.[37] Harvested enterocytes were attached to collagen-coated microcarriers as described previously, and 1.5×10^5 microcarrier-attached enterocytes were injected intraperitoneally into Gunn rats. Control rats were injected with microcarriers alone. After transplantation, rats in both groups were allowed free access to food and water. At 4, 7, 14 and 21 days after transplantation, blood was collected from the tail vein for serum bilirubin determinations. On the same days, rats from each group underwent bile duct cannulation under ether anesthesia for bile collection and analysis for bilirubin glucuronides. Bile pigments were analyzed as nonderivative tetrapyrroles by reverse-phase high pressure liquid chromatography (hplc) using a modification of the method developed by Chowdhury et al.[38] Fifty percent of enterocyte-transplanted Gunn rats demonstrated a significant increase in the concentration of bilirubin monoglucuronide 4 and 7 days after transplantation, and 25% showed a persistent increase 14-21 days after transplantation. Similar results were obtained with bilirubin diglucuronide. No significant differences in serum bilirubin levels between the two groups were noted.

The specific activity of the bilirubin UDPGT in rat small intestine has been shown to be only 10% of the activity of the enzyme in the liver. Based on that observation, it would be expected that the enzyme activity in isolated enterocytes would be less than that seen in isolated hepatocytes. This may explain in part why in a significant number of animals we saw no appreciable increase in bile bilirubin conjugates following enterocyte transplantation. In addition, no other major liver-specific functions have been described in isolated enterocytes. It is thus unlikely that these cells will have any direct utility in the treatment of acute liver failure.

References

1. Sutherland DER, Numata M, Matas AJ et al. Hepatocellular transplantation in acute liver failure. Surgery 1977; 82:124-32.
2. Makowka L, Rotstein LE, Falk RE et al. Reversal of toxic and anoxic induced hepatic failure by syngeneic, allogeneic and xenogeneic hepatocyte transplantation. Surgery 1980; 88:244-53.

3. Baumgartner D, LaPlante-O'Neill PM, Sutherland DER et al. Effects of intrasplenic injection of hepatocytes, hepatocyte fragments and hepatocyte culture supernatants on D-galactosamine-induced liver failure in rats. Eur Sur Res 1983; 15:129-135.
4. Groth GG, Arhborgh B, Bjorken C et al. Connection of hyperbilrubinemia in the glucuronyltransferase deficient rat by intraportal hepatocyte transplantation. Transplant Proc 1977; IX:313-316.
5. Matas AJ, Sutherland DER, Steffes MW et al. Hepatocellular transplantation for metabolic deficiencies: Decrease in plasma bilirubin in Gunn rats. Science 1976; 192:892-894.
6. Vroemen J, Buorman W, Heirwegh K et al. Hepatocyte transplantation for enzyme deficiency disease in congeneic rats. Transplantation 1986; 42:130-135.
7. Demetriou AA, Whiting JF, Feldman D. Replacement of liver function in rats by transplantation of microcarrier-attached hepatocytes. Science 1986; 233:1190-1192.
8. Demetriou AA, Reisner A, Sanchez J et al. Transplantation of microcarrier-attached hepatocytes into 90% partially hepatectomized rats. Hepatology 1988; 8:1006-10090.
9. Holzman MD, Rozga J, Neuzil DF et al. Selective intraportal hepatocyte transplantation in analbuminemic and Gunn rats. Transplantation 1993; 55:1213-1219.
10. Rozga J, Williams RF, Bellew T et al. Transplantation of encapsulated microcarrier-attached hepatocytes into analbuminemic rats. Surg Forum 1991; 42:419-21.
11. Fujioka H, Hunt PJ, Rozga J et al. Carboxyfluorescein (CSFE) labeling of hepatocytes for short-term localization following intraportal transplantation. Cell Transplant 1994; 3:397-408.
12. Rozga J, Holzman M, Fujioka H et al. Repeated intraportal hepatocyte transplantation in rats. Cell Transplantation 1994; 4:237-43.
13. Moscioni AD, Rozga J, Chen S et al. Long-term correction of albumin levels in the Nagase analbuminemic rat: Repopulation of the liver by transplanted normal hepatocytes under a regeneration response. Cell Transplantation 1996; 5:499-503.
14. Eguchi S, Rozga J, Chen SC et al. Treatment of hypercholesterolemia in the Watanabe rabbit using allogeneic hepatocellular transplantation under a regeneration stimulus. Transplantation 1996; 62:1-6.
15. Rosenthal R, Chen SC, Hewitt W et al. Techniques for intrasplenic hepatocyte transplantation in the large animal model. Surg Endosc 1996; 10:1075-1079.
16. Eguchi S, Lilja MH, Kamlot A et al. Loss and recovery of liver regeneration in rats with fulminant hepatic failure. J Surg Res 1997; 72:112-122.
17. Lilja H, Blanc P, Eguchi S et al. Fetal rat hepatocytes: Isolation, characterization and transplantation in the Nagase analbuminemic rats. Transplantation 1997; 64:1240-1248.
18. Arkadopoulos N, Chen S, Khalili TM et al. Transplantation of hepatocytes for prevention of intracranial hypertension in pigs with ischemic liver failure. Cell Transplant 1998; 7:357-363.
19. Rozga J, Jeppson B, Bengmark S. Portal branch ligation. Reevaluation of a model. Am J Pathol 1986; 125:300-308.
20. Neuzil D, Holzman M, Rozga J, Levine J et al. Fibroblast transplantation in rats: Transduction and function of foreign genes. J Surg Res 1993; 54:631-637.
21. Morsiani E, Rozga J, Scott H et al. Automated liver cell processing facilitates large-scale isolation and purification of porcine hepatocytes. ASAIO Journal 1995; 41:155-61.
22. Cai TT, Sun AM. Microencapsulation of recombinant cells: A new delivery system for gene therapy. FASEB Journal 1993; 7:1061-1069.
23. Moscioni AD, Chowdhury JR, Barbour R et al. Human liver cell transplantation. Prolonged function in athymic-Gunn and athymic-analbuminemic hybrid rats. Gastroenterology 1989; 96:1546-1551.
24. Mito M, Kusano M, Onishi T et al. Hepatocellular transplantation. Gastroenterology Jpn 1978; 13:480-490.
25. Gupta S, Aragona E, Vermuru RP et al. Permanent engraftment and function of hepatocytes delivered to the liver: Implications for gene therapy and liver repopulation. Hepatology 1991; 14:144-149.

26. Selden C, Gupta S, Johnstone R et al. The pulmonary vascular bed as a site for implantation of isolated liver cells in inbred rats. Transplantation 1984; 38:812-83.
27. Hamaguchi H, Yamaguchi Y, Goto M et al. Hepatic biliary transport after hepatocyte transplantation in Eizai hyperbilirubinemic rats. Hepatology 1994; 20:220-224.
28. Hicks BA, Drougas J, Arnaout W et al. Impaired liver regeneration in analbuminemic rats. J Surg Res 1989; 46:427-430.
29. Eguchi S, Kamlot A, Ljubimova J et al. Fulminant hepatic failure in rats: Survival and effects on blood chemistry and liver regeneration. Hepatology 1996; 24:1452-1459.
30. Kokudo N, Otsu I, Okazaki T et al. Long-term effects of intrasplenically transplanted adult hepatocytes and fetal liver in hyperbilirubinemic Gunn rats. Transplant Int 1995; 8:262-268.
31. Borel-Rinkes I, Bijma A, Koppers W et al. Evidence of metabolic activity of adult and fetal rat hepatocytes transplanted into solid supports. Transplantation 1992; 2:210-214.
32. Hagihara M, Shimura T, Takebe K et al. Effects of iso and xeno fetal liver fragments transplantation on acute and chronic liver failure in rats. Cell Transplant 1994; 3:283-290.
33. Hoztoglou M, Moorman A, Lamers W. Persistent expression of genes transferred in the fetal rat liver via retroviruses. Som Cell Mol Genet 1995; 4:265-272.
34. Rozga J, Podesta L, LePage E et al. Control of cerebral edema by total hepatectomy and extracorporeal liver support in fulminant hepatic failure. Lancet 1993; 342:898-899.
35. Ringe B, Lubbe N, Kuse E et al. Total hepatectomy and liver transplantation as two stage procedure. Ann Surg 1993; 218:3-9.
36. Hartman F, Bissell DM. Metabolism of the heme and bilirubin in rat and human small intestinal mucosa. J Clin Invest 1982; 70:23-29.
37. Arnaout WS, Moscioni AD, Felcher AH et al. Intraperitoneal transplantation of microcarrier-attached enterocytes in rats. Am J Surg 1989; 157:89-92.
38. Chowdhury JR, Chowdhury NR, Wu G et al. Bilirubin mono- and di-glucuronide formation in human liver in vitro: Assay by high pressure liquid chromatography. Hepatology 1981; 1:622-627.

Index

M

N

O

P

R

S

T

U

V

For Product Safety Concerns and Information please contact our EU representative GPSR@taylorandfrancis.com Taylor & Francis Verlag GmbH, Kaufingerstraße 24, 80331 München, Germany